THE SUSTAINABLE UNIVERSITY

THE SUSTAINABLE UNIVERSITY

Green Goals and New Challenges for Higher Education Leaders

James Martin, James E. Samels
& Associates

Johns Hopkins University Press
Baltimore

Printed in the United States of America on acid-free paper

Johns Hopkins Paperback edition, 2013
9 8 7 6 5 4 3 2 1

Johns Hopkins University Press
2715 North Charles Street
Baltimore, Maryland 21218-4363
www.press.jhu.edu

The Library of Congress has cataloged the hardcover edition of this book as follows:

The sustainable university : green goals and new challenges for higher education leaders / James Martin, James E. Samels & associates.
p. cm.
Includes bibliographical references and index.
ISBN-13: 978-1-4214-0459-2 (hdbk. : alk. paper)
ISBN-10: 1-4214-0459-1 (hdbk. : alk. paper)
1. Universities and colleges—Environmental aspects—United States. 2. College buildings—Energy conservation—United States. 3. Sustainability. I. Martin, James, 1948 Jan. 14– II. Samels, James E.
LB3223.S745 2012
378.1'960973—dc23 2011026544

A catalog record for this book is available from the British Library.

ISBN-13: 978-1-4214-1251-1
ISBN-10: 1-4214-1251-9

Special discounts are available for bulk purchases of this book. For more information, please contact Special Sales at 410-516-6936 or specialsales@press.jhu.edu.

Johns Hopkins University Press uses environmentally friendly book materials, including recycled text paper that is composed of at least 30 percent post-consumer waste, whenever possible.

Contents

Foreword

If one were to conduct a SWOT analysis of the role of sustainability in higher education, one might find that the strengths are limited, while the weaknesses are widespread. The examples in this volume illustrate possibilities that have been realized at a rising number of colleges, community colleges, and universities. Colleges that began looking at how to change their use of energy through conservation technologies and behaviors have found real economic savings while reducing their carbon footprint. Universities have found that "green" buildings not only reduce energy and water costs but also make for a more comfortable working environment. Other schools have gone so far as to become renewable energy producers, anticipating the need to keep both costs and environmental impacts low as far into the future as possible.

Many institutional problems have been created by failures to think about sustainability in both operations and curriculum. Campuses in every region of the country have hundreds of millions of dollars of deferred maintenance that will require either that unprecedented investments must be made to bring them to livable standards or that the buildings be demolished. Those same campuses continue to produce significant amounts of greenhouse gases as they try to heat these failing buildings. Failure to think far enough into the future in planning academic programs and services has resulted in employers' creating sustainability positions for which they have not yet developed adequate resource centers and training programs. The constant struggle among external forces such as creating greater access, demonstrating accountability, and declining public financial support indicates that some institutions will either dramatically change or will disappear over the next decade unless they develop a sound and sustainable fiscal base. Likewise, unsustainable practices that have negative environmental impacts will result in an increase in fines, taxes, lawsuits, and remediation costs. Predicted increases in energy prices will force institutions to look at the economics of unsustainable practices such as the wasteful use of water and poorly designed buildings. Chapter after chapter in this volume forthrightly addresses issues such as sustainable design and architecture, the definition of sustainable academic leadership, and the roles of the president and trustees in these conversations.

Although it is helpful for higher education leaders to be able to assess their current strengths and weaknesses—a process that can be facilitated by the use of the Sustainability Tracking, Assessment, and Rating System (STARS)—it is more important that they be able to look to the future and recognize the opportunities for sustainable practices that will directly advance and strengthen the mission of their institutions. Understanding what should be done to develop a more sustainable college or university is one of the most pressing responsibilities for presidents, provosts, and board members. This volume will be of assistance in raising questions to consider and providing successful action plans to study. Some have argued that when it comes to campus sustainability we have already "picked the low-hanging fruit." Even if this were true, it is a dangerous way to overlook a strategic moment on many campuses. Some institutions have demonstrated that "not-so-low-hanging fruit" are not that difficult to reach when innovative strategies are used. The idea that a campus would replace its fossil-fuel heating plant with a heat-pump geothermal source would have been unthinkable a decade or two ago, but it is now a strategy being enacted in several areas of the country. More than a thousand stories captured in the Association for the Advancement of Sustainability in Higher Education (AASHE) Bulletin over the past several years provide an indication of the many creative ways that campuses have sought to create a more sustainable world. As campuses learn from one another about how to become more sustainable in their practices, they are also learning how to create new degree programs and institutional support services to serve students seeking a sustainable future.

Today the greatest need is for graduates who have the ability and the willingness to approach the vexing problems of society with interdisciplinary experience and systemic thinking. New forms of sustainability education provide not only the context for developing the knowledge and skills to succeed in a complex world but also the platform for implementing a group of high-impact educational programs, including first-year experience, place-based education, and service learning, among others. It may be too simple to describe sustainability education as preparing problem solvers. While that description is accurate, as many chapters in this volume attest, it is also incomplete. A more critical role of sustainability thinking in higher education is meaning making. The phrase "education for sustainability" has been with us for nearly two decades, but it is still not particularly clear to many observers what is meant by it, or even by the term *sustainability*. As we continue to discuss and debate the values, scope, and implications that undergird sustainability, we must "make meaning" of this term for our students, our campus community, and our society. In this regard, Davis Bookhart's chapter on the evolving meanings and perceptions of sustainability over time will prove useful going forward.

Higher education has traditionally played several roles in society, and two are especially important to the purposes of this book. The first is providing graduates with greater opportunities. Sustainability is concerned with providing career and lifestyle opportunities for all in current and future generations, and sustainability education can thus serve a major role in how higher education meets its "opportunity" obligations. A second role that higher education continues to play is the creation of knowledge. In this regard, presidents, provosts, and their faculties critically need to develop deeper and richer understandings of what we mean by sustainability and the creation of a sustainable society. It is sometimes difficult to find the balance between gloom and hope when we work on those issues. Nonetheless, one can be hopeful that higher education leadership teams will use the practical examples provided in the chapters that follow to join in a transformation of how we view ourselves, our relationships to one another, and our environment.

Paul Rowland
Executive Director
Association for the Advancement of Sustainability
in Higher Education
Denver, Colorado

Preface

This book—a guide for presidents, trustees, provosts, vice presidents, senior faculty, and sustainability professionals—offers best practices and action plans to achieve major sustainability goals and objectives on their campuses. While *The Sustainable University* provides a significant amount of fresh advice for many operational areas, one of its chief purposes, apart from fiscal planning per se, is to move beyond individual departments and to offer presidents and board members, in particular, new kinds of benchmarks for the evolving relationship between sustainability thinking and the institution's strategic planning process. As many of the following chapters will demonstrate, sustainability leadership in higher education has become an increasingly complex enterprise; it is now as important an element of a college or university's master planning agenda as its physical plant and maintenance programs.

Presidents' cabinets have been generally underserved by the existing literature on sustainability leadership, much of which focuses on a specialty area such as new campus architecture and LEED (Leadership in Energy and Environmental Design) certification, green curricula, or sustainable residential programs. This book has been designed to address the needs of higher education leaders. It is not a primer on institutional fiscal sustainability, but rather, it is a book that covers executive summaries, benchmarks, and action plans to measure success across a broad set of institutional operations.

Contributors include one of the broadest selections of sustainability professionals assembled in one volume to date, including the founding president of Second Nature, the president of U.S. Partnerships for Education for Sustainable Development (who also serves as co-coordinator of the Higher Education Associations Sustainability Consortium), the current and immediately prior executive directors of the Association for the Advancement of Sustainability in Higher Education (AASHE), the original manager of AASHE's Sustainability Tracking, Assessment and Rating System (STARS), and the senior writer on sustainable architecture and design for the *Chronicle of Higher Education*. These contributors are joined by the current or former directors of sustainability from the following universities: Florida, Colorado, Wake Forest, Harvard, North Carolina, Michigan State, University of California at Santa Cruz, and Johns Hopkins. As well, presidents from Ball State, Wyoming, Lane Community College, and Unity College contribute chapters or extended segments of chapters.

The volume offers several original contributions to the field:

—multiple strategies to institutionalize sustainability thinking successfully among all employees at a college or university (chapter 5)

—five steps to follow in developing multi-institutional partnerships dedicated to sustainable policy objectives (chapter 3)

—the first public summary of the results of AASHE's initial year of the Sustainability Tracking, Assessment and Rating System (STARS) in which more than seventy colleges and universities participated during the 2008–2009 academic year (chapter 4)

—four of the most common myths held by presidents and board chairs regarding commitments to sustainable decision making, along with best practices to overcome them (chapter 1)

—four investing cautions to consider in building a university endowment according to sustainability principles (chapter 12).

Following the foreword by Paul Rowland, executive director of AASHE, and this preface, chapters are grouped into five sections: key trends and new challenges on campuses at present; sustainability and the leadership team; new agendas for key operational areas such as endowment management, campus architecture, and athletics programming; sustainability and external agencies, including accreditation and legal counsel, and a conclusion incorporating principal lessons learned and a set of summary best practices for presidents, provosts, and trustees to achieve campus sustainability objectives.

Part I, "Updating the National Conversation on Sustainability: Key Trends and Challenges," begins with a chapter by James Martin and James E. Samels that focuses on what leaders will learn from the chapters in this volume, how to apply those lessons, and why they are important. In focusing on the numerous ways that sustainability has moved from purely operational advances to guidelines for much of the strategic thinking on campus, the authors present four of the most significant challenges that senior leadership teams face at both public and private institutions in accomplishing a comprehensive sustainability agenda and how to address them. In this process, the authors also examine a group of persistent "myths" about sustainability leadership and decision making and the reasons that they still block progress on many campuses.

In chapter 2, Anthony Cortese, the founding president of Second Nature, one of the two or three most prominent organizations, along with AASHE, attempting to "accelerate movement toward a sustainable future" (Second Nature website, 29 July 2009) for higher education will provide a candid account of the most significant leadership successes—and failures—that he has observed in working with approximately two thousand higher education presidents, trustees, and academic deans in his responsibilities at the organization since 1993.

Debra Rowe, president of the U.S. Partnership for Education for Sustainable Development, and co-coordinator of the Higher Education Associations Sustainability Consortium, with Aurora Lang Winslade, sustainability coordinator at the University of California, Santa Cruz, focuses in chapter 3 on the skills necessary to build sustainability-centered alliances across large numbers of institutions via partnerships and consortia. In the process, they will also highlight lessons and best practices from colleges and universities that have implemented productive, lasting agreements.

In chapter 4, Judy Walton, former executive director of AASHE, and Laura Matson, program manager for the association's STARS program, share the first summary of results from the pilot year (2008–2009) of its "Tracking, Assessment, and Rating" exercise, which included a first group of seventy higher education institutions. AASHE now counts almost nine hundred American colleges and universities in its membership. The results of this new program have been much anticipated and will certainly influence higher education sustainability good practices and policy formulation for the coming decade.

While Rowe and Winslade focused on building sustainability consortia across dozens of institutions, they agree with Leith Sharp, founding director of the Harvard Green Initiative—now called the Harvard University Office for Sustainability—and Cindy Pollock Shea, director of the sustainability office at the University of North Carolina–Chapel Hill, that perhaps the foremost challenge for any campus sustainability coordinator is to institutionalize this thinking throughout an entire institution, and in particular among trustees and senior faculty members who have sometimes been left behind as green initiatives move forward.

Sharp and Shea open chapter 5 with a set of five challenges and action steps facing coordinators attempting to achieve this complex goal. This chapter also features an extended contribution by Michael Baer and Sean Farrell, higher education principals at Isaacson, Miller executive search firm, which outlines six ways the green movement has reshaped both the presidential search process and successful candidates.

As the sustainability movement on some campuses enters its third or fourth decade of development and its themes and strategies are adapted by a new generation of leaders, there is value in assessing the changing meanings of sustainability itself for higher education leaders and their institutions. Davis Bookhart, director of the Sustainability Initiative at Johns Hopkins, comments on the evolving definitions and meanings of sustainability in chapter 6.

Part I closes with a chapter focusing on students by Terry Link, who served as director of the Office of Campus Sustainability at Michigan State University for more than two decades. Link contributes a conversation entitled "Sustainable Citizenship: The Challenge for Students and Their Institutions" that builds directly on the chapters by Rowe and Winslade and Sharp and Shea, as he examines how sustainability professionals and faculty members, especially, can collaborate in preparing "sustainable citizens" on campus who are committed to reshaping society's goals through their career and professional choices.

Part II, "Sustainability and the Leadership Team: New Assignments," opens with a chapter by Jo Ann Gora, president of Ball State University, and Robert Koester, professor and director of Ball State's Center for Energy Research, Education, and Service, that outlines the principal responsibilities for a president personally to accept in leading his or her campus toward sustainable decision making and policy formulation.

In chapter 9, Thomas Buchanan, president of the University of Wyoming, and the university's general counsel, Tara Evans, offer the book's most controversial look at sustainability leadership. Buchanan has been unafraid, during his twenty-seven years at Wyoming, of adopting sometimes unpopular positions regarding fossil fuels, particularly the phenomenon of "clean coal." During his presidency, the Wyoming legislature appropriated $12.1 million to fund the first two years of the university's new School of Energy Resources, an enterprise that immediately laid the groundwork for the Coalbed Natural Gas Center and a Center for Coal Conversion Technologies. Buchanan and Evans discuss the reasoning behind these initiatives as well as focusing on, in their view, some of the unfortunate forms of political correctness that have attached themselves to sustainability thinking on campuses over the past decade.

Mary Spilde, president since 2001 of Lane Community College in Oregon, one of the Northwest's largest higher education institutions, with over 36,000 students, is also the current chair of the American Association of Community Colleges Board of Directors and a board member of the League for Innovation in the Community College. From these complementary perspectives, Spilde presents nine sustainability-focused action steps for institutional leaders that are critical to the two-year college experience and how to implement each through budget, strategic planning, and student life initiatives.

Chapter 11, "Sustainability, Leadership, and the Role of the Chief Academic Officer," is contributed by two chief academic officers and a professor holding an endowed chair, all with extensive experience in academic planning for sustainability at their own institutions and within sustainability-focused associations such as AASHE. Intentionally covering a broad range of institutional types, from small Eco League environmental colleges to major public and private research universities, Geoff Chase (San Diego State University), Peggy Barlett (Emory University), and Rick Fairbanks (Northland College) have developed a detailed overview of the primary roles and responsibilities of academic deans and senior professors in accomplishing sustainability objectives within the academic governance systems.

Part III, "Fresh Agendas for Campus Operations," focuses on leadership goals in five areas: endowment management, architecture and design, campus housing, campus dining programs, and athletics programming and infrastructure. In chapter 12, Mary Jo Maydew, vice president of finance and administration at Mount Holyoke College, focuses on the need to balance short-term financial decisions with their long-term impact as she discusses how to "green an endowment."

Chapter 13, written by Scott Carlson, senior reporter on architecture and design at the *Chronicle of Higher Education,* presents a group of successful projects, accompanied by selected photographs, which have earned high praise for their fresh thinking, cost economies, and sustainable design. Carlson also includes a section on the LEED certification process, comments on why it is worthwhile, and suggests how to prepare an application most effectively. The chapter concludes with a reflection and photograph from President Mitchell Thomashow and Cynthia Thomashow of Unity College on their initial year in the nation's first LEED platinum–certified presidential dwelling.

Norbert Dunkel, Director of Housing and Residence Education at the University of Florida, along with Lynne Deninger, vice president for Cannon Design, writes about the most influential trends in sustainable campus housing in chapter 14. Dunkel also offers a perspective on these trends from his position as recent national president of ACUHO, the Association of College and University Housing Officers.

Rightly and wrongly, dining programs have become the "face" of sustainability on many campuses, as even uninitiated academic administrators learn the vocabulary of "trayless cafeterias" and "reducing institutional food miles." In chapter 15, Howard Sacks, director of the Rural Life Center at Kenyon College, shares the inside story on how Kenyon developed what has become one of the most prominent collegiate dining programs nationally, the Food for Thought Local Food Initiative. As described on the Kenyon website, Food for Thought "educates the public about their food choices, farming, and local rural life," while also keeping millions of dollars in annual food purchases within its county. Sacks also describes in detail the strategic steps Kenyon has taken to achieve 36 percent local foods in its dining halls since 1996.

University athletics programs, described by one sustainability director at the start of this project as one-half of the "holy grail" of higher education (along with university medical centers) provide one of the major influences for or against the institution's sustainability agenda. Dedee DeLongpré Johnston, formerly at Florida and now the director of sustainability at Wake Forest University, along with Dave Newport, director of the Environmental Center at the University of Colorado, undertakes a major assignment in chapter 16 as they describe from the ground level both the successes and the challenges in accomplishing sustainability goals in conjunction with major university athletics. Prominently featured in their narrative will be an informed look at events leading up to the 2009 announcement that the Heavener Football Complex, a $28 million addition to the University's Ben Hill Griffin Stadium, had received LEED Platinum certification, making it the first building of any kind in Florida and the first athletic facility in the nation to achieve this rating. This chapter is accompanied by interviews with two college and university athletics administrators focusing specifically on how their institutions (Middlebury and Yale) achieved significant sustainability goals.

Part IV, "Beyond the Green Gates: Sustainability and the Institution's External Partners," addresses two influences on sustainability beyond the campus: the regional accreditation process, and the impact of legal issues on sustainability initiatives. Sandra Elman, president of the Northwest Commission on Colleges and Universities, is leading the Northwest Commission, among six regional accrediting commissions nationally, toward developing the first formal "green" accreditation standard. Her chapter explores how the sustainability movement is changing the higher education quality assurance process and what steps presidents and provosts can take proactively to prepare their institutions for an increased green awareness in their accreditation relationships.

James E. Samels founded Samels Associates, a higher education law firm, in the early 1980s. In chapter 18, "Green Legal: Creating a Culture of Vigilance, Compliance, and Sustainability Thinking," he and James Martin define the most important legal issues that have accompanied the rise of sustainability in higher education and offer legal strategies to address continuing challenges in this area.

The volume's concluding section, Part V, "The Complex Path Ahead," is a chapter by Martin and Samels that offers a group of summary lessons drawn from preceding chapters for presidents, deans, and board members, as well as a final list of eight best practices for management teams implementing sustainability leadership.

The authors would like to thank several individuals for helping to make this book possible. First, Anthony Cortese and Debra Rowe, at the center of the national sustainability conversation since its inception, offered pivotal advice at early stages of the project to move it onto a focused and solid footing. Along the way, several campus leaders in particular, within a larger group of many generous contributors, made a difference through their insights: Davis Bookhart at the Johns Hopkins University, Geoff Chase at San Diego State University, Norbert Dunkel at the University of Florida, Jo Ann Gora at Ball State University, and Cindy Pollock Shea at the University of North Carolina, Chapel Hill. Thanks also to Paul Rowland, Judy Walton, and Laura Matson at AASHE for joining the project on first ask during an extremely busy pilot year for the STARS program.

We appreciate once again the significant contributions of Catherine Jacquet, both for her editorial expertise and for her unfailing good cheer in copy editing the entire manuscript, her third higher education title with us for the Johns Hopkins University Press. Thanks also go to Arlene Lieberman, Esq., of Samels Associates, Attorneys at Law, and Kate McLaren, post-doctoral research fellow at The Education Alliance, for their key contributions and research for chapter 18. We also want to recognize the continuing support and encouragement of Ashleigh McKown, the press's former higher education editor, and to thank her for the early confirmation that such a book would be useful to many presidents, provosts, and trustees. Sadly, we would also like to take this moment to remember the work and legacy of Rick Fairbanks, former provost of Northland College and coauthor of chapter 11. During the completion of that narrative, Rick became ill, yet he completed his sections on the role of the chief academic officer with characteristic grace and candor before passing in February 2011. All involved in the project extend their sympathies and appreciation to Rick's wife, Angela, and their daughters.

Finally, we dedicate this book to the student consumers about to enter higher education institutions. As they construct virtual futures via Facebook and Twitter, we hope that they extend and enhance the hard work of organizations like Second Nature and AASHE when they choose their careers and that they bring to those careers as much imagination and courage as did those leaders who have moved sustainability to this point in our history.

I

Updating the National Conversation on Sustainability: Key Trends and Challenges

CHAPTER ONE

The Sustainable University: A Need to Move Forward

James Martin and James E. Samels

Sustainability Leadership and Its New Priorities

Once a set of basic environmental guidelines for various operational managers on campuses, sustainability leadership has become something far more powerful and complex over the past generation in higher education. Sustainability itself, when translated into college and university terms, is now about setting priorities, achieving organizational transformation, and undertaking cultural change. This volume has been designed to serve as a comprehensive yet pragmatic resource for three groups of higher education leaders, in particular, who have received little help to date for these tasks: presidents, provosts, and trustees.

In this regard, *The Sustainable University: Green Goals and New Challenges for Higher Education Leaders* is aimed at the leadership teams of research universities, liberal arts and career colleges, and community colleges in providing action plans, best practices, emerging trends, and an examination of some of the most persistent "myths" about sustainability and higher education leadership. The main priority of all contributing authors is to provide these underserved leaders with a clear sense of the tools they need, how to apply them, and why this process is so important to their campuses in the increasingly connected world of global higher education.

For a chief executive officer or a trustee committee chair, sustainability is now as much about strategy as operations, as much about broad institutional identities as individual position papers. Why, then, have these leadership teams continued to be underserved by the literature of sustainability? Anthony Cortese, founding president of Second Nature, believes that to this point studies of sustainability and higher education leadership, particularly those few chapters aimed at presidents or board members, have fallen short because writers

> continue to define sustainability as simply *environmental, scientific, or technological,* rather than as an element of the core mission of higher education: to produce graduates who will shape a thriving, civil society. Presidents, trustees, and many academic vice presidents and deans are now realizing that generations of undergraduates have come and gone with very little understanding of the importance of aligning their personal and professional lives with sustainability principles, much less the action steps to accomplish this. Sustainability is increasingly strategic, and perhaps the most persistent wrong that presidents and provosts must now address is that sustainability initiatives are viewed as "off to the side" when strategic plans are developed for their campuses.[1]

This dynamic needs to change, and the chapters in this book have been developed to move beyond prior studies and to provide these senior administrators and trustees with specific plans to achieve these goals.

Geoff Chase, dean of undergraduate studies at San Diego State University and chair of the board for the Association for the Advancement of Sustainability in Higher Education (AASHE), observes that, for their part, chief academic officers balance multiple commitments, and the literature on sustainability has not caught up with their evolving responsibilities: "It has not adequately addressed the role that provosts and deans can play as leaders of the institution's academic program in the context of sustainability. What is now needed is a broad discussion of how chief academic officers can work more efficiently with presidents, their faculties, and administrative colleagues to advance the sustainability agenda."[2] Chase and his colleagues Peggy Barlett from Emory University and Rick Fairbanks from Northland College contribute this kind of broad conversation about sustainability and the chief academic officer's role in chapter 11 of this volume.

Davis Bookhart, director of sustainability at the Johns Hopkins University, undertakes an even longer view of sustainability and its challenges for contemporary leadership teams in chapter 6, "Sustainability: Shifting Definitions and Evolving Meanings." Bookhart believes that sustainability, at its core, is about transformation, going from what we have today to what we hope will be tomorrow. However, the fact that sustainability has grown in scope and purpose so quickly over the past two decades forms much of the challenge for a president or provost when a simple yes or no decision can implicate significant resources for a period of several years at least. As Bookhart explains, positive impressions of a sustainable future do not always translate clearly into simple action steps. In response, the leaders of many universities and colleges are caught searching for solutions that are not too radical for the board and not too conservative for the current sustainability director and next year's student consumers. The chapters in this volume provide examples of this kind of solution along with strategies to link research and practice more effectively.

Our research and interviews have identified the following four challenges as among the most formidable facing presidents committed to advancing sustainability goals on their campuses. We also provide several successful responses to these challenges and conclude the chapter by examining and dismissing some of the most persistent "myths" about sustainability and higher education leadership that can compromise the efforts of even innovative chief executives, deans, and trustees.

Four Challenges Facing Leadership Teams

1. *To institutionalize sustainability thinking.* Following approximately two hundred conversations and interviews with sustainability professionals in the preparation of this volume, this challenge emerged as one of the two or three most serious on a majority of their campuses. Leith Sharp and Cindy Pollock Shea, two experienced sustainability coordinators from Harvard University and the University of North Carolina–Chapel Hill, respectively, focus in depth in chapter 5 on the significant difficulties presidents, provosts, and especially campus directors of sustainability face as they move their agendas forward. Anthony Cortese confirms these challenges with his comment: "The cultural operating instructions of modern society are that if we just work a little harder and smarter and let the market forces do their thing, all these challenges will work themselves out." However, he quickly adds, "*We have a de facto systems design failure.* The 21st century challenges must

Four Challenges for Institutional Leaders

- To institutionalize sustainability thinking
- To develop and implement an effective system of sustainability benchmarks
- To implement a flexible and accountable budget model to support sustainability goals
- To engage trustees in the campus sustainability agenda

be addressed in a systemic, integrated, and holistic fashion. Achieving sustainability calls on us to have a deliberate societal strategy to make a rapid transition to a low carbon, less auto dependent, and circular production economy."[3]

A key factor to recognize is that, as Geoff Chase adds, "Sustainability is inherently interdisciplinary. It is about connectedness and interdependence, and the structures of our institutions do not often reflect that kind of cross-boundary thinking. Disciplinary boundaries . . . [can be] major hurdles."[4] Cortese and Chase agree that sustainability initiatives suffer from the silos that have grown up at universities, liberal arts colleges, and community colleges over the past half-century as resources are not shared, personnel are stretched too thin, and institutional governance routinely lags behind key planning, funding, and construction timelines.

Debra Rowe, president of the U.S. Partnership for Education for Sustainable Development, and Aurora Lang Winslade, sustainability manager at University of California–Santa Cruz, acknowledge the silo effect while suggesting finely tuned people skills—and patience—to get objectives accomplished: "Interpersonal skills, sometimes called 'soft skills' are extremely important because sustainability crosses traditional hierarchical, class, cultural, and disciplinary boundaries and requires collaboration, cooperation, and inspiration from the entire community."[5]

Whatever the skill sets necessary for lasting change, there are several action steps most campuses must take to overcome this challenge. As described by Ann Rappaport and Sarah Hammond Creighton, authors of *Degrees That Matter: Climate Change and the University,* they include the following:

—The campus master plan includes sustainability

—Energy management systems are in place

—Funding for energy efficiency is provided

—Standards are in place for new construction

—No old lighting technology is on campus

—A recycling program is in place.[6]

Students have become increasingly articulate in identifying some of the largest challenges to institutionalizing sustainability thinking. At Northland College, a founding member of the Eco League of environmental colleges, a group of undergraduates posted the following in their alumni magazine: "Community is the key. Northland is still learning this lesson. Like every institution, it has inertia, a resistance to change. . . . But at the heart of every effort there are individuals painstakingly advancing the College, pushing us toward our written goals and ideals. . . . Every uphill inch we've climbed has been earned through human effort."[7]

2. *To develop and implement an effective system of sustainability benchmarks.* While there persists among some sustainability professionals the unspoken belief that self-regulation is cheaper and just as effective as external benchmarking systems, this challenge is being gradually met and overcome by the new Sustainability Tracking, Assessment and Rating System (STARS) program introduced and then piloted in 2008–2009 by the Association for the Advancement of Sustainability in Higher Education. AASHE has been careful to take the time to develop a new kind of rating system, rather than a purely competitive ranking model such as those being formulated by journals such as *Forbes* and the Princeton Review. As Judy Walton, former acting executive director of AASHE and now its director of membership and outreach, explains, "A rating system provides a clear road map for a campus to reach a benchmark level at any time. In contrast, a ranking system provides no clear target."[8]

For STARS, colleges and universities will be assessed in three categories: Education and Research, Operations, and Administration and Finance; participation is voluntary, and institutional reports will be posted on a website hosted by AASHE and open to the public. Ratings will apply for three-year periods.[9] As one example of a campus-based concern about the challenges of implementing a comprehensive rating system that can avoid the "horse race dynamics" of competitive, if misconstrued, institutional ranking systems,[10] Gioia Thompson, University of Vermont's director of sustainability, comments, "We've all been using different metrics and emissions factors, and it's hard to compare with lots of variables."[11]

In treating sustainability achievements as components of a rating rather than a ranking system, AASHE has significantly helped current higher education leaders move beyond simple "to do" lists for sustainability success to "connecting the dots and grasping the larger patterns. . . . Driving STARS participants to adopt a whole systems approach [elevating] sustainability beyond individual projects."[12]

3. *To implement a flexible and accountable budget model to support sustainability goals.* Campus leaders, especially presidents, CFOs, and trustee finance committee members, will not travel too far down any path

related to sustainability until its associated costs are identified and thoroughly assessed. Unfortunately, these considerations often take place within confusing and only partially informed contexts. Budgeting strategically to accomplish sustainable objectives is one of the most fraught challenges for the leadership team. Optimally, there needs to be in place a budget process that can flexibly address the larger institution's sustainability needs and priorities at any point in a fiscal year. Norb Dunkel, director of housing and residence education at the University of Florida and coauthor of chapter 14, describes the dilemma facing CFOs: "Even the most willing higher education administrators have concerns regarding the identification of funds to advance sustainable initiatives. The up-front costs of these initiatives can take away from long overdue maintenance and renovation projects. There is still not a full business plan understanding of the return on investment."[13]

Rowe and Winslade confirm the presence of these disconnects within most budgeting models:

> It is not simply a lack of funding; it is also the way that budgets are organized that impedes success. A problem encountered on almost every campus is the separation between capital and operating budgets. True savings, both financial and in reducing climate impact, are realized when up-front investments in building greener, more energy efficient buildings are made. Often small, up-front costs are value engineered out in the design phase as limited capital budgets require cuts. Those making the cuts are frequently not aware of the cost over the long-term life of the building, and they have no direct incentive to reduce those costs. . . . Another example . . . is the fact that on most campuses, energy bills are paid out of central funds not by individual units. Again, this removes the financial incentives for users to reduce energy use.[14]

Added to this, the entire budget-building experience for sustainability on many campuses is now held to stricter forms of scrutiny and accountability. New stakeholders playing a role in developing the academic budget can include suppliers, regulators, activist groups, student consumers, and the employers of undergraduates and recent graduates,[15] as CFOs can be forced to choose between being conservative stewards of the campus and approving the latest "politically sexy, expensive program that may have little impact on greenhouse gases and reducing our carbon footprint," as Robert Weygand, vice president for administration and chief finance officer for the University of Rhode Island acknowledges. Weygand, also serving as chair of the university's Council on Sustainability, continues, "My greatest fear is that we will devote our efforts toward sustainability initiatives that are more in vogue than in the science of improving our environmental standing. There is little or no discretionary spending left on campuses. Sustainability programs realize that they must be financially integral to the University's mission if they are to be included." One recommendation that has worked at URI has been to design budget requests that link directly with programming in sustainability-related areas that can then draw "larger groups of new students to our campus by combining minors in sustainability with graduate options such as our joint oceanography and MBA program."[16]

4. *To engage trustees in the campus sustainability agenda.* Even without the challenges of implementing significant sustainability initiatives on their campuses, it is not an easy time to be a college or university trustee. The national association for trustees, the Association of Governing Boards of Universities and Colleges, spends a good deal of its trustee orientation and professional development agenda specifically on the leadership and financial fundamentals of being an effective citizen volunteer. Issues such as the trustees' stake in sustainable decision making are described by one AGB commentator as a more "complex and nuanced realm."[17] Still, whether a trustee committee is actively considering sustainability concerns or not, the fact remains that most board members need an extensive orientation to sustainability on their campuses, and not simply for its budget impact, in order to govern effectively.

Stephen Pelletier, in his *Trusteeship* article, "Sustainability: What is the Trustee's Stake?" contends, "For trustees, the takeaway message . . . is that sustainability has become a central consideration across the board in university *business* decisions. Passing a tipping point, it has become a pivotal focus of management." And, if their institution happens not to be talking about these topics, board members should assess "when, where, and how" sustainability appears in strategic planning documents as well as beginning to assess the financial implications of sustainability from up-front costs to expected returns down the road.[18]

San Diego State orients all of its new trustees each year to help them understand that sustainability is not an entirely new field in higher education but, rather, "a view of our current work on campus—teaching, research, service—through a new and important lens."[19] Institutions that engage their trustees in regular orientations and provide timely updates on sustainability programs position themselves for success over the long term, particularly when it becomes clear that funds beyond one budget cycle will be required to complete major initiatives related to new construction, endowment development, or faculty salary upgrades.

These are four of the most significant challenges that presidents, trustees, academic vice presidents, and chief finance officers face when undertaking sustainability-focused projects that require major budget allocations, faculty and administrator time commitments, and institutional marketing energies. A group of persistent myths and misconceptions about sustainability leadership further complicate these challenges and influence decision making by senior officers.

Myths about Sustainability and Higher Education Leadership

We have written a column entitled "Future Shock" for *University Business* magazine for almost a decade, and during the writing of this book, we discussed in that column a number of "green myths" related to sustainability and higher education leadership that have emerged. These misconceptions, along with the challenges just noted, form a counterpoint to the best practices for institutional leaders that appear in the concluding chapter of this study.

1. *Sustainability is about institutional operations.* Sustainability is without question about institutional operations, but the misconception held by more than a few college and university leaders, especially trustees, is that it is only about operations. Laura Matson, coauthor of chapter 4, "Measuring Campus Sustainability Performance: Implementing the First Sustainability Tracking, Assessment, and Rating System (STARS)," offers a candid response: "The largest misconception on campus is that sustainability leadership is all about the campus's energy consumption. Energy conservation is hugely important . . . so it is no wonder that many institutions start key initiatives here, but energy conservation is not the only thing that colleges can do to demonstrate sustainability leadership. Sustainability in higher education encompasses more than just operations, and, even within operations per se, schools can demonstrate creative leadership through other activities, including waste reduction, environmentally preferable purchasing, and the elimination of toxins."[20]

Four Myths about Sustainability and Higher Education Leadership

- Sustainability is about institutional operations.
- Sustainability raises (or lowers) costs.
- There is a common path to sustainability goals.
- LEED is the answer.

In his presentation of "Higher Education Modeling Sustainability as a Fully Integrated Community" in 2008, Anthony Cortese presents university operations as one of four core elements constituting a fully integrated community, also including education, research, and relations with local communities. Cortese believes that institutional operations contribute one aspect of a more "complex web of experience and learning" critical to achieving transformative change.[21] Looking beyond individual campuses, Mitchell Thomashow, president of Unity College in Maine, cautions that sustainability must be much more broadly defined so that even higher education constituents are reminded that it is primarily "a response to a planetary challenge involving climate destabilization and species extinction."[22]

While sustainability does affect all areas of institutional operations, to judge it as solely this is to be blinded to its larger priorities, as the author of the provocatively titled *Wired* magazine article, "It's Not Just Carbon, Stupid," exhorts: "Tackling climate change is vital. But to see everything through the lens of short-term C02 reductions, letting our obsession with carbon blind us to the bigger picture, is to court catastrophe. Climate change is not a discrete issue; it's a symptom of larger problems."[23]

2. *Sustainability raises (or lowers) costs.* This misconception has passionate adherents on both sides. Viewed from either direction, the cost factor discussion produces a cohort of leaders who believe either that sustainability-driven decisions must be taken because they save funds, often a great deal of funds, over the long term, or that even though these decisions do not save funds, they still must be taken because the overarching goals of sustainability trump the need for short-term budget reserves.

LEED (Leadership in Energy and Environmental Design) certification decisions convey the dilemma just described. Allowing key sustainability goals at the institution to be reduced to the money question time after time becomes problematic. Scott Carlson, senior writer for the *Chronicle of Higher Education* and author of chapter 13, argues that a considerable number of presidents and senior administrators have held back from applying for a certification that carries immediate prestige and marketing advantages simply because of its perceived price: "Some institutions decide that the cost and the hassle are too much. . . . Some architects and builders estimate the price of certification at $1000 to $2000 per point, which has added up to more than $100,000 for some colleges. An official at the U.S. Green Building Council says that the cost is usually much lower—$12,000 to $22,000—and that it will continue to fall as more consultants learn more about green building. The Council pockets just $450 to $600 per registered building."[24]

One way to move beyond this misconception is offered by Michael Baer, vice president and director of the higher education search practice at Isaacson, Miller, as he reflects on how the executive search process at his firm now addresses links between sustainability awareness and financial expertise among final candidates: "We frequently encounter individuals on campuses who believe that the major reason for encouraging sustainability is for economic reasons and that it will reduce costs. It follows in their minds that business operations are the most critical, such as facilities and food services, rather than creating a broader awareness of the long-term social benefits to be achieved in educating students, even the entire campus population, about the environmental gains from the green movement."[25] Baer advocates that successful candidates arrive without preconceptions regarding sustainability costs, no matter what their prior experiences, that they evaluate carefully the institution's ongoing links between mission and resource commitments, and, finally, that they avoid tendencies to stereotype sustainability initiatives either as revenue-drains or cost-reducers.

3. *There is a common path to sustainability goals.* Over the course of approximately two hundred conversations and interviews for this book, it became apparent that while many believe that the period for orientations to sustainability has passed and the movement has transitioned from its "awareness-building phase" to a "solutions-building phase,"[26] there is still a pressure, especially among administrators who came to a sustainability agenda late in their careers, to think conventionally and seek similar solutions. Add to this the emergence of high-profile national benchmarking systems, and the "right thing to do" qualities of sustainable policies and practices can exert a palpable influence on campus thought and constrain risk-taking and innovation.

As one reporter in the popular press noted, "A surfeit of buzz words can obscure the real meaning of sustainability. . . . Check Google for the phrase 'sustainable development' and 26 definitions pop up, littered with buzz words such as 'preservation,' 'eco-system,' 'biological system,' 'resource base,' and 'social equity.' " He then asks, "Is there any developer these days who does not tag 'sustainable' onto the description of 'executive homes' or a condominium development?"[27] As a counterpoint, Laura Matson encourages university and college leaders not to seek standard solutions to current challenges but rather to build on the natural diversity of opinion in higher education in designing programs to fulfill individual institutional missions: "At some institutions, students are the major drivers of sustainability while others struggle to engage their student bodies. Likewise, institutions' abilities to source local food, generate clean, renewable energy on site, or get employees to commute using alternative methods are influenced significantly by their local contexts. That said, institutions are also working collaboratively to overcome those local constraints by advocating for sustainability policies locally and nationally."[28]

As many institutions undertake mission statements that incorporate a general sustainability commitment along with specific goals, there is still the belief among leaders of the Presidents' Climate Commitment and AASHE, for example, that new models to achieve these objectives can and should be tried—and will succeed. The chief architect at the Illinois Historic Preservation Agency summarizes this belief with a simple phrase: "We're just *stealth green*. We don't show it—we have no solar panels, no collectors, no whiz-bang things. We're taking old buildings and putting them back in use and making them more green."[29] In sum, there is no common path to sustainability objectives, and experienced sustainability directors help their institutions address this ambiguity by supporting, yet also challenging, their colleagues to use the time and resources necessary to define those paths forward that are best for their college or university.

4. *LEED is the answer.* There is perhaps nothing within the sustainability movement in American higher education that sparks as much controversy, competitiveness, and misconceptions as the U.S. Green Building Council's LEED certification process. As Scott Carlson explains, "In higher education, where sustainability is a hot issue, LEED certification is often a visible symbol of a college's commitment. Since LEED began in 2000, more than 1500 college projects have been registered . . . but some college officials are raising questions about the process of LEED certification. Some say it emphasizes less-important priorities in building. Others believe the certification is costly and a pain."[30]

As the sustainability movement matures, successful LEED campaigns are becoming more powerful indicators of influence and prestige. At the same time, a misconception has emerged that one LEED building can push your institution to the next level of prominence regionally, if not nationally. As Mitchell Thomashow, president of Unity College and someone who has blogged for the past year about his experiences living in a LEED Platinum–certified home on campus, cautions, "A major myth would be to believe that one showcase LEED building solves the problem."[31]

Concerns and suspicions such as these are influencing multi-hundred-million-dollar infrastructure decisions across the nation's 4,100 higher education institutions. Stanford University, planning a $400 million science and engineering quadrangle had doubts about LEED because it did not go far enough in recognizing some green building features that the university was considering. Colorado State University at Fort Collins plans to match the LEED gold standard in a $400-million construction project but not to pursue certification because the process, by its rules, could cost up to $100,000. Those funds will instead be dedicated to additional green features in the eventual structure. It should be added that both of these universities are also planning simultaneous construction projects on their campuses that will seek LEED certifications.[32] The LEED "brand" has thus become powerful enough to produce significantly divergent opinions on the same campus as to its value.

Robert J. Koester, director of the Center for Energy Research, Education, and Service at Ball State University, comments in Carlson's article that some university administrators now expect a payback on green building elements that they would never expect from their own endowment: "'Frequently, value engineering is couched from the point of view of years' payback,' with college administrators taking a pessimistic view of energy-saving features that pay off in the long run."[33] Another Colorado-based sustainability administrator, David Newport, director of the Environmental Center at the University of Colorado–Boulder and coauthor of chapter 16, summarizes a final LEED concern: "As for largest misconceptions, one must be that you can 'green-build' your way to carbon neutrality. It does not matter how good your buildings are—unless they are zero carbon buildings, of which there are few, if any—you cannot grow campus size and get to zero. Efficiency will not get us to zero. . . . Bottom line: conservation rules."[34]

Conclusion: A Need to Move Forward

Joel Garreau, long-time *Washington Post* journalist and newly-named Lincoln Professor of Law, Culture, and Values at Arizona State University, opens his most recent book, *Radical Evolution,* with a simple assessment, "Innovation arrives more rapidly than does change in culture and values."[35] Many presidents and trustees are now operating within unpredictable and unresolved economic environments as their campuses advance into deep considerations of high-cost sustainability program and infrastructure innovations for which the price tags are outpacing even optimistic enrollment and budget projections. Combine these tensions with a few quietly voiced suspicions by board members that "green" thinking may be a fad that will fade after a few more graduating classes, and it becomes clearer how important pragmatic leadership resources for senior officers will be going forward. For example, a group of architects spoke out at a Higher Education Buildings Conference in London in December 2009 to question some universities' "professed environmental zeal," warning them that their efforts to improve their environmental credentials were often "unsustainable and misguided." Rod McAllister, partner in Sheppard Robson, a 70-year-old London design firm whose website declares its guiding principles as "innovation, sustainability, and social conscience," complained, "I don't buy it. A lot of it is just greenwash. There's a lot of tearing down buildings and putting up highly efficient new ones, when it is much better to adapt existing structures."[36]

When opinions as provocative as this earn an international forum, it is certain that some trustees and presidents will be influenced. Yet in response, the methods to institutionalize sustainability thinking articulated in the chapter by Sharp and Shea, the partnering principles put forth in Rowe and Winslade's discussion, and the specific campus mobilization action steps provided by Gora and Koester in their chapter on the presidency, to name three, have been designed, like this book as a whole, to articulate a useful balance of viewpoints while helping campuses to move forward with practical, affordable solutions. Some of the most important long-term goals for leadership teams still remain simple ones, as a *Wall Street Journal* reporter explains: "Although recycling is important, it isn't as effective as reducing the use of materials from the get-go."[37] Or, as David Newport succinctly adds, "Efficiency won't get us to zero. Zero will get us to zero. We can't add 'efficient energy load' and expect the load to go down."[38] A larger point for institutional leaders, however, is that their next steps will almost uniformly need to be broader and bolder ones in terms of mission clarity, budget allocations, and personnel assignments.

As a corollary, some of these ambiguities are being addressed in the corporate world by the arrival of high profile "eco-officers." The *Los Angeles Times* reports, "As companies grapple with climate change, try to attract eco-conscious customers and develop alternative energy agendas while complying with regulations, a new kind of administrator is moving into the executive suite to help out."[39] While higher education has seen a steady rise of sustainability directors and coordinators over the past decade, one wonders how many of them would readily acknowledge the constraints in naming a single individual on campus to this position if the rest of the institution has not grasped, and accepted, its share of leadership for change.

For all of the progress even enlightened colleges and universities have made to date, sustainability continues to evolve in its meanings and influence on campuses and in the communities that surround them. As new challenges and opportunities continue to emerge, higher education institutions are being forced to redefine and address new sustainability goals—and to challenge their presidents, trustees, and senior leadership teams to guide them to greater levels of engagement and effectiveness.

CHAPTER TWO

Promises Made and Promises Lost: A Candid Assessment of Higher Education Leadership and the Sustainability Agenda

Anthony D. Cortese

American higher education has been granted tax-free status, the ability to receive public and private funds, and academic freedom in exchange for educating students and producing the knowledge that will result in a thriving civil society. Each year our higher education institutions enroll more than eighteen million students and produce more than three million graduates into society and its workforce. Community colleges, colleges, and universities have been a critical leverage point in making a modern advanced civilization possible for an unprecedented number of people in almost every important way. American society continues to see higher education as a national resource and crucial to the development of economic success, a view that was originally put forth by the report by President Truman's Commission on Higher Education in 1947. In addition, society looks to higher education to solve current problems, anticipate future challenges, develop innovative solutions, and model the behaviors that society must adopt to continue to evolve in a positive direction.

Humanity and Higher Education at a Crossroads

American higher education now faces a challenge larger and more severe than any it has ever addressed. Through the exponential growth of our population and its technological and economic systems, humans have become the dominant influence on the health and wellbeing of the earth. The sum of humanity and the expansive dynamic of industrial capitalism constitute a *planetary* force comparable in disruptive power to the Ice Ages and the asteroid collisions that previously redirected Earth's history. While the earth's population has grown from one billion to 6.7 billion in the last two centuries, energy consumption has risen eighty times and economic output has risen sixty-eight times, most of this over the last half century.[1] Despite the impressive array of environmental protection laws and programs in industrialized countries since 1970, all living systems—oceans, fisheries, forests, grasslands, soils, coral reefs, wetlands—are in long-term decline and are declining at an accelerating rate, according to most major national and international scientific assessments. Humans are challenged by a staggering array of toxic chemicals that persist and are affecting our health and the viability of large ecosystems.

At the same time, we are not succeeding in many health and social goals: more than three billion people are without sanitation and earn less than $2.50 a day, and more than a billion have no access to clean drinking water. Worldwide, the gap between the richest 20 percent and the poorest 20 percent has jumped from 30:1 to 78:1 in one generation.[2] Even in the United States, the gap is the greatest since the late nineteenth century. As 25 percent of the world's population consumes more than 70 percent of its resources, we are beginning to experience wars over oil and water that are destabilizing society. Beyond this, a challenge that will accelerate all these negative trends is human-induced global warming, primarily from the burning of fossil fuels, that is now destabilizing the earth's climate and most of its other life-supporting systems. Despite what may be read in the news media, especially in the United States, human-induced climate disruption is real and is already affecting us faster than predicted by the most conservative scientists in 2007. What most people do not understand is that destabilizing the earth's climate can undermine modern civilization. As Dianne Dumanoski asserts in her 2009 book, *The End of the Long Summer,*

> Our way of life depends on a stable climate. The cores of ice drilled from the ice sheets on Greenland and Antarctica tell us we live at a truly extraordinary time within the Earth's volatile climate history. Through most of our species' 200,000-year existence, our ancestors had to cope with a chaotic climate marked by extreme variability, a climate that could not support agriculture. The world as we know it, with agriculture, civilization, and dense human numbers, has only been possible because of a rare interlude of climatic grace—a "long summer" of unusual climatic stability over the past 11,700 years. The human enterprise has become a risky agent of *global change*. The gargantuan size of our modern industrial civilization is now disrupting our planet's very metabolism—the vast overarching process that maintains all of earthly life. Because of humanity's planetary impact, this exceptional moment on Earth is drawing to a close. What lies ahead is a time of radical uncertainty.[3]

The crucial question for our society is: How will we ensure that *all* current and future humans will have their basic needs met and live in thriving, secure communities in a world with nine billion people that plans to increase economic output by four to five times by 2050. This is no longer simply an environmental challenge; rather, we now need to rethink and remake the human presence on earth through what I have described as the aspiration of a sustainable society.

The Need for a Change in Mindset

The routine business of our civilization is threatening its own survival. We continue to be guided by myths of human domination of nature and of continuing "progress" fueled by economic growth because they have worked over the last three centuries to create a modern society offering spectacular increases in the quality of life for a significant portion, though still a minority, of the world's population. The guiding myth assumes that human technological innovation will allow us to ignore planetary limits. Moreover, we are dominated by linear short-term thinking that makes it difficult to recognize the cumulative dangers of current actions or to see that the impact of collective humanity is now global, intergenerational, and prone to rapid, unexpected shifts. For example, greenhouse gas emissions released today will begin to have their most serious effects in thirty to fifty years and will continue for several centuries. In Western industrialized society, many still view increasing material consumption as the principal measure of individual and group success, despite increasingly negative health, social, and environmental effects.

Through economic globalization, we are spreading this cultural paradigm even while it makes our societies more vulnerable to the growing instability of natural and human systems—witness the impacts of the 2007–2008 failure of the housing and financial markets that continue to depress the worldwide economy; the 2010 volcanic eruption in Iceland that wreaked havoc with travel to and from Europe and with the worldwide economy; and, even more recently, the oil spill in the Gulf of Mexico that focused immediate attention on who was responsible rather than on the long-term consequences of this disaster and the urgent need to reduce fossil-fuel sources in our lives for environmental and social justice issues. We still view this array of health, economic, energy, political, security, and environmental issues as separate, competing, and hierarchical when they are actually *systemic* and *interdependent. We have a de facto systems design failure.* These challenges must be addressed in an integrated and holistic fashion with an emphasis on creating new and more desirable ways of helping society succeed, such as local sustainable food production that provides healthy food and local jobs and that protects soils and water supplies.

We must also confront the reality that our current higher education system is reinforcing the unhealthy, inequitable, and unsustainable paths that society is pursuing. As David Orr has said, "It is not a problem *in* education; it is a problem *of* education."[4] The structure of higher education reinforces many of the cultural assumptions referred to above. The guiding myths of humans as separate from nature and nature as primarily a source of resources to be utilized and controlled for human purposes are prevalent paradigms in higher education.

As one example of the many undesirable consequences of the current educational paradigm, our ecological, health, and social footprint is largely *invisible* to most of us and almost completely absent in the price of products. Our economic system acts as if the price is right for all products, but in truth their prices are *mostly wrong* because they do not reflect the negative impact on human health and communities and on the earth's ecosystems. Currently, the price is the proverbial tip of the cost iceberg. The best estimates of the true life-cycle health, social, and ecological cost of a gallon of gasoline, for example, is between $8 and $12. That cost does not reflect the ecological damage of oil spills or the destabilization of the earth's climate. As a result, the average American does not realize that through our economic system, we consume the equivalent of our body weight in solid materials daily, over 94 percent of which goes to waste before we ever see the product or the service. For example, it takes a few thousand pounds of material to make a laptop computer weighing five to six pounds. Consequently, the market fails in efficiently allocating resources and allows us to continue in a group self-deception about the impact of our daily living. A key assignment for college and university curricula and teaching methods is thus to make invisible impacts visible.

Culture and Values before Scientific Development

How can we create a healthy, just, and sustainable society? Let us imagine that all current and future generations are able to pursue meaningful work and have the opportunity to realize their full human potential both personally and socially. Imagine that communities are strong and vibrant because they celebrate cultural diversity, are designed to encourage collaboration and participation in governance, and emphasize the quality of life over the consumption of materials. In an August 2007 *New York Times* story, "In Silicon Valley, Millionaires Who Don't Feel Rich," executives with a net worth of more than $5 million discussed how their families were unhappy with their lifestyle because they compared themselves with neighbors whose net worth is five to ten times greater. As one executive complained, "Here, the top 1 percent chases the top one-tenth of 1 percent, and the top one-tenth of 1 percent chases the top one-one-hundredth of 1 percent. You try not to get caught up in it," he added, "but it's hard not to."[5] The road to sustainability is one of culture and values as much as it is about scientific and technological development. It must be guided by the arts, humanities, social and behavioral sciences, and religion as much as by the physical and natural sciences and engineering.

Can this be done? A growing consensus of business, government, labor, and thought leaders now believe that a "clean, green economy" based on these principles is the best way to restore American economic leadership, create millions of jobs, help solve global health and environmental problems, and recreate geopolitical stability and justice. For example, DuPont has reduced heat-trapping emissions by 72 percent since 1990 and saved $3 billion.[6] Ray Anderson, chairman and founder of Interface, Inc., the world's largest modular carpet manufacturer, with annual sales of $1.2 billion, and one of the world's leading companies dedicated to economic, social, and ecological sustainability says:

> At Interface, the business case for sustainability (as a core purpose of our business) is crystal clear: A capitalist to the core, I can't think of a better business case than lower costs, better products, higher morale, loyal employees and goodwill in the marketplace. Our costs are down, not up, dispelling the myth that sustainability is expensive. Our first initiative—zero-tolerance waste—has netted us $433 million in saved or avoided costs, more than paying for all capital investments and other costs associated with sustainability. Our products are the best they've ever been. Sustainability is a wellspring of innovation; our product designers have been particularly successful using "biomimicry" as a guide, nature as inspiration. Our people are galvanized around our mission and a shared higher purpose—Maslow at his best: self-actualization that comes when people commit to something bigger than themselves, a type of top-to-bottom and bottom-to-top alignment that sustainability has fostered. The goodwill of the marketplace is tremendous, winning business for Interface because customers want to be aligned with a company that is trying to do the right thing. No amount of marketing, no clever advertising campaign, could have created the kind of customer loyalty that we have experienced.[7]

Thinking like Anderson's represents a key societal shift. There have been six major economic downturns in the last fifty years. In the first five of these, many in industry, state legislatures, and Congress called for relaxing environmental, health, and safety standards to cope with the economic challenges. In the current one, the opposite is occurring. Environmentally preferable actions are now viewed as the best way to restore and sustain economic stability.

The Role of Colleges and Universities in Sustainability Leadership

As presidents, provosts, and trustees, in particular, assume leadership roles in helping to make this a reality, their institutions will operate as fully integrated communities that model social, economic, and biological sustainability, along with interdependence with their local, regional, and global environments. In many cases, we think of teaching, research, operations, and relations with local communities as separate activities; they are not. In fact, these activities form a flexible network of experience and learning since all operational segments of a college or university system are critical to achieving this *transformative* change. Practically speaking, five action steps need to occur, starting with chief executive and chief academic officers:

1. *The content of learning* will reflect interdisciplinary systems thinking, dynamics, and analysis for all majors and disciplines with the same *lateral rigor* across, as the *vertical rigor* within, the disciplines.

2. *The context of learning* will change to make human and environmental interdependence, values, and ethics a seamless core of teaching in all the disciplines rather than isolated in special courses or modules.

3. *The process of education* will emphasize active, experiential, inquiry-based learning and real-world problem solving both on campus and in the larger community.

4. Higher education will *practice and model sustainability*. A campus will "practice what it preaches" and model economically and environmentally sustainable practices in its operations, planning, facility design, purchasing, and investments and will link these efforts to the formal curriculum on an ongoing basis.

5. Finally, institutions will implement *new forms of partnership* with their local and regional communities to help make them socially vibrant, economically secure, and environmentally sustainable. Universities and colleges have the ability to create new and better markets for goods and services that will improve society in all ways, not just in narrow economic terms.

Frank Rhodes, former president of Cornell University, suggests that the concept of sustainability offers "a new foundation for the liberal arts and sciences." It provides a new focus, sense of urgency, and curricular coherence at a time of drift, fragmentation, and insularity in higher education, what he calls "a new kind of global map."[8] In these ways, sustainability leadership provides a vital source of hope and opportunity to facilitate institutional renewal and to revitalize higher education's sense of mission.

Current Realities for Presidents and Others

Over the past two decades, unprecedented growth has occurred in distinct academic programs related to the *environmental dimension* of sustainability in higher education. Innovative environmental, and now sustainability, studies and graduate programs are rapidly growing in every major scientific, engineering, social science, business, law, and religious discipline on campuses. Progress on campuses modeling sustainability has grown at an even faster rate. Higher education has embraced programs for energy and water conservation, renewable energy, waste minimization and recycling, green buildings and purchasing, alternative transportation, local and organic food growing and "sustainable" purchasing. According to the U.S. Green Building Council, the higher education sector has nearly four thousand new buildings that are being designed or have been designed to meet advanced levels of sustainable design under the LEED system (Leadership in Energy and Environmental Design) simply since 2000.[9] The weekly bulletin of the Association for the Advancement of Sustainability in Higher Education (AASHE), the primary national network of colleges and universities working on sustainability efforts with over one thousand members, routinely cites fifty to sixty new sustainability initiatives in all aspects of campus life. The student sustainability and environmental movement is clearly one of the most well-organized and sophisticated student initiatives since the anti-war movement of the 1960s. Collectively, these developments represent one of most encouraging trends in higher education innovation since World War II.

Unfortunately, college and university leaders, and even senior faculty members, are performing less effectively on the health, social, and economic dimensions of sustainability. The overwhelming majority of graduates know little about the importance of sustainability or how to lead their personal and professional lives aligned with sustainability principles. This is largely a result of the deep and often hidden cultural assumptions described earlier, the disciplinary dominated educational model, the separation of classroom learning from the application and practice of sustainable living on campus, and an increasing emphasis on education for commerce and career. Critically important educational innovations about civic engagement and community service, inquiry-based and experiential learning, international perspectives, and environmental stewardship have been developed over the past two decades, and they are still not sufficient to make the transformative shift necessary at the rate and scale needed. Nor are the important efforts of nongovernmental organizations such as Second Nature and the Campaign for Environmental Literacy, of student organizations like the Energy Action Coalition, and of fundamental higher education professional associations such as the AASHE Consortium going to prove adequate in accomplishing necessary goals. On balance, these efforts still have not been significantly integrated with other socially focused movements such as civic engagement, social justice, economic development in impoverished parts of the United States and the world, and human rights. With a few exceptions, sustainability writ large, as suggested in the opening sections of this chapter, is not a central institutional goal or lens for determining the success of higher education institutions. It is largely viewed as one of many priorities that will be handled as time and resources allow, and as such it does not represent a challenge to the existing purpose or structure of higher education.

The Challenge for Campus Leadership Teams

In addressing sustainability decisions, presidents, provosts, chief finance officers, and trustees still view many of the challenges as environmental, not societal, and not as fundamental to how they can meet the basic needs of all current and future humans in a fair, equitable, peaceful, and sustainable manner. Most higher education administrators and faculty members do not understand the urgency with which society must begin to reform the way it is operating and the extent to which their curricula need to focus on social, economic, and ecological sustainability in order to fulfill their obligation to society. There is acceptance by presidents, chief academic officers, and many members that the current education system as it is designed and practiced has led to the unhealthy, inequitable, and unsustainable path that society is pursuing. The institutional rewards and incentives and the largest amounts of external funding for education and research continue to favor disciplinary over interdisciplinary or trans-disciplinary models, making it difficult to address the systemic and interdependent challenges of society.

The current higher education structure largely leaves students to integrate many different and often conflicting ways of viewing the world without adequate tools for systemic analysis and integration. Senior administrators, especially in highly rated four-year colleges and research universities, are often reluctant to engage with faculty over educational direction because of the resistance of faculty to administrative influence on academic freedom. *These are systemic challenges.* As a scholar of higher education, Louis Menand, has observed,

> One thing about systems, especially systems as old as American higher education, is that that people grow unconscious of them. The system gets internalized. It becomes a mind-set. It is just "the way things are," and it can be hard to recover the reasons *why* it is the way things are. When academic problems appear intractable, it is often because an underlying systemic element is responsible, but no one quite sees what or where. People who work in the academy, like people in any institution or profession, are socialized to operate in certain ways, and when they are called upon to alter their practices, they sometimes lack the compass to guide them.[10]

Innovation by addition and growth over the past generation has not challenged colleges and universities to ask hard questions about whether the current structure is fulfilling its mission. For the first time since World War II, higher education is now faced with a number of serious *structural* constraints in financial resources that are not likely to be alleviated in the foreseeable future. This is very unfamiliar territory for many presidents and board members. Moreover, the change in mindset needed to face the gathering challenges that society faces is prompting a large number of people to question whether higher education can make the necessary systemic transformation far enough or fast enough without strong outside influence. In *Universities and the Future of America,* former Harvard president Derek Bok outlines several concerns:

> When society recognizes a need that can be satisfied through advanced education or research *and* when sufficient funds are available to pay the cost, American universities respond in exemplary fashion. . . . On the other hand, when social needs are not clearly recognized and backed by adequate financial support, higher education has often failed to respond as effectively as it might, even to some of the most important challenges facing America. . . . After a major social problem has been recognized, universities will usually continue to respond weakly unless outside support is available and the subjects involved command prestige in academic circles.[11]

While it is unclear whether a majority of higher education institutions can or will lead this transformation, there are some recent examples of success in sustainability-driven thinking and action that can offer a road map for leadership teams.

Systemic Change: How Some Have Achieved It

The American College & University Presidents' Climate Commitment

One of the brightest beacons of light for systemic change in U.S. higher education has been the American College & University Presidents' Climate Commitment (ACUPCC), launched in February of 2007 by twelve college and university presidents, working with Second Nature, the Association for the Advancement of Sustainability in Higher Education (AASHE), and ecoAmerica.[12] The ACUPCC is a high-visibility, joint and individual commitment to measure, reduce, and eventually neutralize campus greenhouse gas emissions, to develop the capability of students to help larger society do the same and, most importantly, to report publicly on their progress. Three years later, as of July 2010, 673 colleges and universities, covering all fifty states and the District of Columbia, have made this commitment. They represent 5.9 million students—about 35 percent of the national student population—and include all types of institutions, from community colleges to the largest research universities. The ACUPCC is an example of unprecedented leadership by college and university presidents and their institutions for several reasons.

Higher education is the first major national sector with a significant number of its members to commit to climate neutrality. This is especially important, given the slow pace of the international community and the U.S. Congress to act. The participating presidents believe that leading society to a low-carbon, less auto-dependent economy complements and enhances the educational, research, and public service missions of higher education.

The ACUPCC represents the first time since World War II that higher education has taken on a major societal challenge without prompting or funding from outside sources. The action by these presidents, chancellors, and provosts is sending a strong signal to society that climate change and other large-scale unsustainable and societal practices are real, that urgent action is needed, and that higher education is taking tangible steps to model sustainable behavior and to provide the knowledge and educated graduates necessary for society to do the same.

In pledging to become climate neutral campuses, these leaders are pledging to do what is scientifically necessary, not what is easily achievable, within their current systems of operation. The presidents realize that it will be extremely difficult to achieve climate neutrality in the next 20–30 years, and they are challenging their institutions to achieve this goal because the best scientific research, often produced by their own colleges and universities, indicates that it is necessary.

The positive impact of collective leadership by a large number of college and university leaders will be felt worldwide. Creating a sustainable society is a global challenge requiring solutions of immense proportions. The scale and speed of this challenge demand an unprecedented level of collaboration among leadership teams, boards of trustees, faculty, and students because actions by individual institutions will not be sufficient.

The ACUPCC has accelerated efforts to integrate academic, research, operational, and community outreach actions in a holistic approach to sustainability. There is substantial anecdotal evidence on how effective the commitment has been in raising the importance of all sustainability initiatives on campuses. According to presidents, chancellors, and provosts at numerous participating colleges and universities, the ACUPCC has done as much to create and sustain a vibrant community and sense of shared purpose across those institutions as any other initiative in their recent histories.

The ACUPCC has fundamentally shifted higher education's attention on sustainability from a series of individual program efforts to a broader strategic imperative among chief executive and chief academic officers, business officers, and board members. Sustainability is becoming a key lens for measuring the success of the institution as a whole.

New Majors in Sustainability Studies

The number of institution-wide efforts to ensure that sustainability is a foundation of all learning and practice is small but growing. There are currently a large number of degree-granting programs for environmental, natural resource management, and sustainability specialists at all levels of American higher education. Over a thousand colleges and universities now offer an undergraduate interdisciplinary degree program in some form of environmental or sustainability studies. These programs, which reach an important segment of the student population, are key to ensuring that business, government, and NGOs will have the specific kinds of expertise necessary to address increasingly complex sustainability challenges directly.

Sustainability-Focused Institutional Missions

A smaller number of private and a few public baccalaureate-granting liberal arts colleges have either been created or have evolved to make sustainability the *core mission* of their education and practice. They emphasize interdisciplinary learning, experiential learning on campus and in their local communities, and model sustainable action in their institutional operations. Notable institutional examples include College of the Atlantic and Unity College in Maine, Green Mountain College in Vermont, Northland College in Wisconsin, Prescott College in Arizona, Alaska Pacific University, Berea College in Kentucky, Warren Wilson College in North Carolina, and the Evergreen State College in Washington.

A Bold Experiment: Arizona State University

One of the biggest challenges is in convincing a large public or private research university to make sustainability a core goal of research and education because of its size, the decentralized nature of its decision making and operations, and the academic independence of the individual schools as well as their strong connections to external funding institutions and professional communities.

Arizona State University, based in Tempe, with 67,000 students and nearly 18,000 faculty and staff members, began an experiment in 2002 to change the dominant paradigm of the institution. Michael Crow, its new president, set out to create what he calls the "New American University." One year later, he established what has become the Global Institute of Sustainability (GIOS), and in 2004 ASU founded the full degree-granting School of Sustainability, which has steadily grown to over six hundred undergraduate majors and nearly one hundred graduate students. In addition, all first-year and transfer undergraduates, totaling over 20,000, are exposed to principles of sustainability through the required "ASU 101" course. James Buizer, whom President Crow recruited from the National Oceanic and Atmospheric Administration to lead the conceptualization, formation, and design of GIOS indicates that "injecting a sustainability program into a university requires courageous leadership and commitment throughout the institution, beginning at the very top, and including through to the college deans, school directors and departmental chairs. Without the vision and active leadership from President Crow, GIOS would never have happened."[13]

Over almost a decade, this concept of the New American University has transformed ASU, guided by eight "design aspirations": (1) Leverage Our Place; (2) Transform Society; (3) Value Entrepreneurship; (4) Conduct Use-Inspired Research; (5) Enable Student Success; (6) Fuse Intellectual Disciplines; (7) Be Socially Embedded; and (8) Engage Globally. Each of these aspirations is critical to creating a sustainable society, and together they hold great promise as a template for other engaged universities and colleges.

ASU has been aggressive, experimental, and entrepreneurial in working to fulfill its commitment to sustainability. Recognizing the limits presented by the traditional disciplinary structure, President Crow explained progress to date in the May/June 2010 issue of *Trusteeship:* "More than 20 new transdisciplinary schools, including such entities as the School of Human Evolution and Social Change and the School of Earth and Space Exploration, complement large-scale initiatives such as the Global Institute of Sustainability (GIOS) and the Biodesign Institute, a large-scale, multidisciplinary research center dedicated to innovation in healthcare, energy and the environment, and national security. In the process, we have eliminated a number of traditional academic departments, including biology, sociology, anthropology, and geology."[14]

This approach represents what is probably the most aggressive and explicit attempt to address the invisible mindset and the systemic challenges in higher education discussed previously, although ASU admits that it still struggles with cultural challenges related to traditional reward structures, the expectations related to tenure and promotion, and the tensions inherent in convincing multiple researchers that "use-inspired" knowledge creation is of equal value. Still, in the view of Joel Garreau, author of *Radical Evolution,* who became ASU's Lincoln Professor of Law, Culture and Values in 2009, "Arizona State University has become the nation's foremost silo-busting institution. As one example, across the street from my office is Biodesign in which groups of people are quietly working on inventing our futures with 400,000 square feet of creatures that do not exist in nature. President Crow has done an amazing job of rewiring the DNA of the University—especially in such a short time. Cross-disciplinary collaboration has become so much the norm that one tends to forget it is not like many other places, until one is reminded what it was like in ages past."[15]

Conclusion: Campus Leadership as Sustainability Leadership

While there has been rich scholarship on leadership in higher education, to date there has been little research examining the complex elements within sustainability leadership. One newly published study by Glenn Cummings, deputy assistant secretary of education in the U.S. Department of Education, posed a key question: What common characteristics and actions were taken by successful university and college leaders in pursuit of sustainability? Four institutions of higher education that have been considered national leaders on sustainability were chosen for detailed analysis: two four-year public research universities, Arizona State University and University of New Hampshire; and two two-year public community colleges, Cape Cod Community College and Foothill D'Anza Community College. The author found many common traits and strategies among the leaders:

Leadership plays a crucial role in the success of sustainability implementation. The four presidents realized these central factors: (a) higher education is positioned centrally to help society move on a more sustainable path; (b) sustainability success requires a comprehensive institutional shift; (c) presidents, most of all, hold a unique role in conveying the importance of an institutional commitment to sustainability; (d) sustainability needs to become stated as an aspect of the institution's mission; and (e) sustainability provides an opportunity to reshape each institution to meet its highest ideals.

Administrative policies, particularly in human resource management, can create significant long-term focus on sustainability. To varying degrees all four institutions attempted to influence hiring, and in some cases decisions on tenure and promotion, to support the movement toward sustainability. They also created complementary practices, expectations, and norms to reinforce sustainability.

Effective leaders use the power of "milestones" to underscore the institutional significance of sustainability. Presidents and provosts used important events to strengthen institutional pride and embed sustainability thinking and goals into the school's "brand." At UNH, for example, the entire 2008 commencement ceremony centered on the theme of sustainability, including the first "turning of the switch" of the new methane pipeline that provides over 80 percent of the campus energy needs.

Leaders can use their ability to "tell the story" of sustainability to engage funders in providing new resources to the institution. Three of the four schools achieved national prominence in sustainability partly or wholly as a result of private philanthropy or government grants. The presidents' and trustees' skill in making the case with external funders and both internal and external stakeholders was key to their success.[16]

These examples and others are also discussed in a recent publication by the ACUPCC entitled "Leading Profound Change."[17] They reflect an understanding that for higher education leaders to move their institutions in this direction requires four major overarching perspectives. First, for colleges and universities to fulfill their roles related to sustainability, transformative institutional strategies are required. Second, all effective strategies toward these ends must be communicated to and endorsed by everyone in the campus community. Third, success depends on a dedicated group within the community empowered to lead the process over the long term. And finally, provosts and other leadership team members must hold all involved accountable for success and measure their outcomes. Assessment, in this instance, must go far beyond merely "checking the box" reporting that sustainability is a priority. It will require patient, persistent attention and the kind of stamina and vision to inspire large numbers of people to imagine a better future and collaborate in creating it.

Some higher education observers continue to argue that achieving climate neutrality and sustainability as a society and convincing college and university administrators to lead this effort is impossible. However, if we continue business as usual, today's students and their children will experience much harsher effects of climate disruption and unsustainable means of meeting human needs, finding themselves caught in a world with greatly diminished prospects for quality of life, peace, and security. The challenge for boards of trustees and leadership teams is one of both courage and mobilization. Toward this end, Richard Cook, who retired in 2008 as president of Allegheny College and is one of the founders of ACUPCC, wrote to a colleague who had not yet made the commitment:

> I liken this pledge to President Kennedy's promise to get men to the moon and back within the decade. Neither he nor a cadre of engineers and scientists knew exactly how this would be accomplished or if, indeed it could be. But making a bold pledge to accomplish a strategically important end spurred attention, resources, talent, and urgency to a lofty goal that would be difficult to attain. In much the same way, the Commitment to becoming climate neutral institutions will spur development and accountability, and will surely, in most cases, produce more and better results in a shorter period of time than something short of a specific target. The collective voice of higher education can spotlight our sincere concern and commitment to action in ways that few if any other sectors can. We have largely provided the research that has highlighted the climate concern; we also can provide many of the solutions. If the colleges and universities don't lead, who will?[18]

The goals are outlined before us. Some will follow, and some must lead.

CHAPTER THREE

Trends, Skills, and Strategies to Catalyze Sustainability across Institutions

Debra Rowe and Aurora Lang Winslade

Sustainability is an updated, more accurate understanding of the world that can help us all make smarter decisions. Our educational systems have so far not succeeded in teaching the skills, knowledge, and attitudes necessary to create a sustainable society of healthy ecosystems, social systems, and economies. As individuals and as societies, we have not acted on numerous opportunities to reduce human suffering and improve quality of life while meeting basic human needs in a sustainable manner. We can learn healthier ways of interacting with our environment and each other, and due to the urgency of the challenges we face, systemic actions that shift our institutional, community, national, and global systems toward sustainability rapidly and deeply are now necessary.

Employers continually ask colleges and universities to produce graduates who can work effectively in groups and solve real-world problems.[1] We know a lot about how to develop healthier relationships and interpersonal skills, yet this is rarely part of the core curricula. While economic scarcity is taught in many basic higher education economics programs, nowhere in a standard undergraduate curriculum do we explore how to create sustainable abundance. How different would society be if relationship skills, financial literacy, discussions and planning for sustainable abundance, and problem solving that builds change-agent skills were routinely included and valued university curricula? How different would our institutions be if, during their undergraduate years, students and staff used campuses as living laboratories for sustainable technology and as places to practice new behaviors for a sustainable society?

The good news is that many colleges and universities are asking these questions and responding with a variety of strategies. Many institutions are leading the way toward a sustainable future through changes to the physical campus and its policies and practices, integrating sustainability into all degree programs, and forming local and regional partnerships to improve the economy and the health of ecosystems.

This chapter provides an overview of some of the national trends in sustainability in higher education and offers resources to help catalyze and amplify sustainability initiatives. We then present a selection of tested strategies currently being used by sustainability advocates on campuses. Finally, we offer a profile of the essential skills needed by leadership teams to accomplish long-range sustainability objectives. In this chapter, our goal is to help presidents and trustees, in particular, to understand the transformations now taking place in American higher education and to use this information to create more effective funding and management of the sustainability programs and services at their institutions.

National Trends and Resources to Transform Campuses

Institutional leadership teams are now realizing not only that sustainability cannot be ignored but also that it represents one of the most valuable teaching and learning opportunities on their campuses. These leaders understand that the implementation of sustainability is fundamental to their colleges because it is a smart business practice and because it is critical to their institutional missions of teaching, research, and public service. Many are taking bold action to address the integration of sustainability into all levels of campus life. This immediately involves crossing traditional barriers of disciplinary and departmental silos and organizational hierarchies.[2] Higher education is now reorienting itself to include sustainability in mission and planning, curriculum, research, facilities and operations, student life, purchasing and finance, community partnerships, and alumni outreach. Governance structures are adapting to these shifts as committees, positions, and policies are modified or newly established. Colleges and universities are also partnering with state and national politicians to develop sustainability-oriented legislation and to raise public awareness concerning the benefits of sustainability initiatives.

The Value of Mission and Planning

Colleges and universities are completing new climate protection and sustainability plans as well as integrating sustainability thinking into existing master planning efforts and institutional statements of vision, values, and mission. In the coming decade, we will gradually see more and more colleges signing regional and national commitments, establishing sustainability policies with public statements, and releasing declarations.

Multiple best practice examples are available on the website of the Association for the Advancement of Sustainability in Higher Education (AASHE). AASHE also hosts a listing of more than eighty institutions that incorporate sustainability in their strategic and master plans, including Dickinson, Oberlin, Georgia Tech, California State University at Chico, University of Vermont, Carnegie Mellon, and Green River Community College. The Society for College and University Planning (SCUP) has Sustainability Fellows and regularly offers sustainability-related professional development for its members and the broader community.[3]

Sustainability and Institutional Governance Structures

The ten-campus University of California system, one of the first university systems to pass a comprehensive system-wide sustainability policy in 2003, has approximately sixteen campus-wide sustainability officers. Across the country, campuses are steadily establishing offices of sustainability and sustainability task forces and hiring sustainability directors. These individuals are charged with conducting sustainability assessments, writing campus-wide sustainability plans, and catalyzing and implementing new programs at all levels of the college or university. In a 2008 survey, AASHE found that 90 percent of sustainability officers' positions were created within the last ten years, most of those within the last five.[4] The AASHE "Campus Sustainability Officers Page" has a wealth of resources, including a directory, surveys, and job descriptions. As presidents and deans grapple with the implications of sustainability at their colleges and universities, they are finding a strategic solution in streamlining their internal governance structures, where possible, to accommodate the growing impact and importance of sustainability goals. They see the value of having experts who can help create cultures of sustainability, target cost savings and efficiencies, and guide the university to adopt best practices in a rapidly changing environment.

The Role of Curriculum Development and Internships

Among chief academic officers and faculty leaders, there is a growing recognition that all students need to learn how to engage in sustainability challenges and solutions. In their roles as consumers, investors, family members, and community members, it is centrally important for tomorrow's leaders to integrate sustainability thinking into their daily lives. College graduates are increasingly required by potential employers to understand basic sustainability principles. A survey of more than 1,300 business professionals found that 85 percent of respondents saw environment and sustainability knowledge as valuable, particularly in new hires, while 78 percent said that knowledge would rise in importance over the next five years.[5] As discussed in more detail in chapter 11 on curriculum and academic leadership by Geoff Chase, Peggy Barlett, and Rick Fairbanks, colleges are addressing this by creating new sustainability minors, majors, and concentrations. An increasing number of institutions have included sustainability learning outcomes as part of the general education core curricula and throughout a large number of disciplines. These learning outcomes often include both information about sustainability and the intrapersonal and interpersonal skills to be systemic thinkers and effective change agents.[6] In traditional academic style, curricula are often including materials that encourage students to become literate about sustainability challenges and insisting on critical thinking about ecosystems, poverty reduction, social equity, health, and economic issues. In addition, in an emerging national trend, students are being given opportunities to learn and apply practical problem-solving and interpersonal skills to real world sustainability issues in their assignments. Specific examples of these curricula and internships are provided in chapter 11.

As noted in related chapters by Sharp and Shea, Link, and Gora and Koester in particular, the proportion of today's students that are provided with learning opportunities to think and act in alignment with sustainability principles, whatever their career choices, is steadily rising. An increasing number of campuses from all types of colleges, including private, public, premier research institutions, large state schools, small liberal arts, and community colleges, are modeling sustainability. Professional associations for presidents, such as AACC (American Association of Community Colleges) and AASCU (American Association of State Colleges and Universities), have created sustainability task forces. AACC's Sustainability Education and Economic Development (SEED) online resource center is compiling and sharing information on solar energy, wind power, energy efficiency, green building, sustainability education, and other topics as they relate to innovative practices and partnerships to enhance local economies and stimulate new career pathways.

The Importance of Financing Sustainability Effectively

More presidents and chief finance officers are viewing budget allocations for sustainability as an investment that pays off directly—through cost savings and cost avoidance—and indirectly through enhanced image and community relations, student recruitment, and the capacity to comply with evolving environmental policy requirements. Robert Franek, senior vice president and publisher of the 2010 Princeton Review College Hopes & Worries Survey, reported: "Students and their parents are becoming more and more interested in learning about and attending colleges and universities that practice, teach and support environmental responsibility. According to our recent College Hopes & Worries Survey, 64 percent of college applicants and their parents said having information about a school's commitment to the environment would impact their decision to apply to or attend it."[7]

The author of *Financing Sustainability on Campus,* published by the National Association of College and University Business Officers (NACUBO), outlines strategies for funding projects from both internal and external sources. Among those described are revolving loan funds, gifts, endowments, alumni giving, grants, rebates, incentives, and energy hedges.[8] There is such a strong understanding that investing in sustainability is worthwhile that almost all sustainability officer positions are paid for through internal general funds.[9] A common and essential strategy that should not be overlooked is using savings from energy efficiency projects to reinvest in funding sustainability initiatives.[10] At the University of British Columbia, an institutional trendsetter in sustainability, "the decision to reinvest operational savings from sustainability measures into UBC's sustainability program was key to UBC's ability to move forward with strong and successful measures."[11]

Despite all of this progress, sustainability professionals and programs still struggle against some trustees' and senior administrators' perceptions that sustainability is simply one more competing concern, at risk of being sidelined when budgets are tight. This is a danger that should not be underestimated, especially in light of the effects of the 2008–2009 economic crisis, even though many chief finance officers are learning that the benefits of investing in sustainability mean lower risk via less price volatility, more cost effective savings, and an enhanced ability to attract students.[12]

A Summary of Key National Networks and Resource Sites

Several national networks made up of mainstream higher education associations and institutions now support new sustainability-focused professional standards. They document an expanding body of knowledge and resources about best practices and implementation tools for campus sustainability.

AASHE annually hosts a national conference that draws more than a thousand professionals, publishes a weekly newsletter, and maintains an unmatched resource center of innovations, case studies, and events. In addition to AASHE, other notable higher education networks and organizations include Second Nature, the Higher Education Associations Sustainability Consortium (HEASC), and the Disciplinary Associations Network for Sustainability (DANS).

Second Nature, a national nonprofit, provides support to higher education leaders to integrate sustainability into practice and learning. The organization helped create and supports the groundbreaking American College & University Presidents' Climate Commitment (ACUPCC). Second Nature catalyzes national initiatives and runs capacity-building programs such as the Advancing Green Building in Higher Education program that helps address green building challenges in under-resourced institutions with a three-year grant from the Kresge Foundation.[13] They also support the growing engagement of higher education leaders in federal policy. The ACUPCC is an important network in its own right. As Anthony Cortese, president and co-founder of Second Nature, explains, "The ACUPCC is . . . the only collective initiative of higher education institutions (now 673 institutions in all 50 states and the D.C., representing 35% of the student population) that commits to *three firm goals* over time—becoming climate neutral campuses, ensuring that all graduates have the knowledge to help all society do the same, and publicly reporting on their progress. This is an unprecedented commitment, especially because Higher Education is the first major U.S. Sector with a significant number of its members to commit to climate neutrality."[14]

The Higher Education Associations Sustainability Consortium (HEASC) is "a network of higher education associations with a commitment to advancing sustainability in both their constituencies and in the system of higher education itself. HEASC helps higher education exert strong leadership in making education, research, and practice for a sustainable society a reality. Higher education's leadership is critical to help businesses become sustainable; to have strong thriving and secure communities; and to provide economic opportunities for the broadest number of people while preserving the life support system on which all current and future generations depend."[15] More than fifteen mainstream and national higher education professional associations are members of HEASC, serving college and university presidents, business officers, facilities and housing directors, student affairs staff, planners, recreational managers, event planners, trustees, and procurement officers (see table 3.1).

The organization supports various national initiatives such as ACUPCC and AASHE's STARS, and member organizations have developed national conference calls and webinars, informed policy and the media, shared best practices, built websites on sustainability, catalyzed national task forces, incorporated sustainability into professional development and conferences, and published a variety of resources. The resources from the HEASC members have helped to set precedents and catalyze sustainability efforts on hundreds of campuses. As just a sample,

Table 3.1. HEASC Member Organizations

AACC	American Association of Community Colleges
AASCU	American Association of State Colleges & Universities
AASHE	Association for the Advancement of Sustainability in Higher Education
ACCED-I	Association of Collegiate Conference & Events Directors—International
ACE	American Council on Education
ACPA	College Student Educators International
ACUHO-I	Association of College & University Housing Officers International
AGB	Association of Governing Boards of Universities & Colleges
APPA	Association of Higher Education Facilities Officers
CCCU	Council of Christian Colleges & Universities
NACA	National Association for Campus Activities
NACUBO	National Association of College & University Business Officers
NAEP	National Association of Educational Procurement
NAICU	National Association of Independent Colleges & Universities
NIRSA	National Intramural-Recreational Sports Association
SCUP	Society for College & University Planning

HEASC organizations have contributed sustainability lists of learning outcomes and assessments, student and residential life activities, funding and financing opportunities, operational strategies and resources, and presidential and trustee leadership resources. Their efforts are creating new professional norms for their members and making sustainability literacy and engagement in system-shifting solutions part of the identity for their respective professions.[16]

More than twenty national academic disciplinary associations have initiatives for sustainable development as part of a national network called the Disciplinary Associations Network for Sustainability (DANS) (see table 3.2). Members and staff work to infuse sustainability into curricula and the promotion, tenure, and accreditation processes as well as provide support to integrate cross-disciplinary approaches and to identify new funding opportunities. DANS has brought experts to the Media Strategies for Sustainability project that describes how to change behaviors toward social and environmental responsibility and how to

Table 3.2. DANS Member Organizations

DANS Member Organizations
American Academy of Religion
American Anthropological Association
American Association for the Advancement of Science
American Association of Colleges for Teacher Education
American Chemical Society
American Institute of Biological Sciences
American Marketing Association
American Philosophical Association
American Political Science Association
American Psychological Association
American Society for Engineering Education
American Society of Civil Engineers
American Society of Mechanical Engineers
American Sociological Association
American Studies Association
Aspen Institute
Association for Environmental Studies and Sciences
Association for the Advancement of Sustainability in Higher Education
Association for the Study of Literature and the Environment
Association of American Colleges and Universities
Association of Collegiate Schools of Architecture
Association of University Leaders for a Sustainable Future
Association to Advance Collegiate Schools of Business International
Broadcast Education Association
College Student Educators International
Computing Research Association
Ecological Society of America
International Society for Ecological Economics
Mathematical Association of America
National Academies, Division on Earth and Life Studies
National Association of Biology Teachers
National Council for Science and the Environment
National Humanities Alliance
National Women's Studies Association
Society for College and University Planning
U.S. Partnership for Education for Sustainable Development
U.S. Society for Ecological Economics

This information is available at www2.aashe.org/dans/associations.php.

effectively share information about higher education sustainability efforts with the community. DANS members also encourage publishers to work with faculty to include sustainability principles, themes, and examples in the revisions of textbooks in all disciplines.

Colleges and universities around the world are exploring how to integrate society's growing understanding of the nature of sustainable systems into all aspects of the institution. Each of the organizations and networks described here hosts a broad menu of resources that offer support and guidance to both administrators and faculty members seeking to set and achieve targets for campus sustainability.

New Strategies for Presidents and Leadership Teams

It is through the *integration* of teaching, research, operations, planning, co-curricular activities, community outreach, and public service that colleges and universities most effectively realize their commitment to educating tomorrow's leaders and moving to a more sustainable society. Students learn as much from how a university operates and relates to the local community—often referred to as the "shadow curriculum"—as from what is said and assigned in the classroom.[17] Making explicit the often hidden impacts of the university's consumption and investment practices and providing action opportunities for students provides social, economic, and environmental benefits, more efficient business processes, and enhanced learning. All of the strategies and skills outlined in this chapter are designed with this commitment to integration in mind.

In chapter 2, "Promises Made and Promises Lost: A Candid Assessment of Higher Education Leadership and the Sustainability Agenda," Anthony Cortese articulates the magnitude of the challenges humanity faces and the underlying myths that have generated our unsustainable

Five Strategies to Integrate Sustainability into Higher Education Institutions

- Understand the political: build relationships and coalitions, manage the emotional climate, and count your votes to get the job done.
- Build an effective communication strategy.
- Create systemic changes in the curricula.
- Institutionalize sustainability.
- Eliminate systems that are barriers to sustainability, particularly regulatory and financial barriers.

systems, identifying the need for transformational institutional strategies. How does an institution achieve the changes required? What are the strategies? How does broad acceptance come about? We suggest five effective strategies to integrate sustainability into higher education institutions.

1. *Understand the politics: Build relationships and coalitions, manage the emotional climate, and count your votes to get the job done.*

How can campus sustainability advocates build critical mass for effective sustainability transitions? First, sustainability coalitions can be designed to include the disparate factions on campus that are interested in different parts of sustainability education. Diversity leads to greater understanding among groups and therefore social health, one of the key components of sustainability. Students or faculty members might already be mobilized around civic engagement, global education, or learning communities. Find out who the decision makers are, recognize and connect with them, connect them with one another, and honor the work already being done, especially by those groups who may not yet self-identify as part of sustainability. By demonstrating appreciation both in public and private for the efforts of a variety of people, larger coalitions of support can be built. Giving credit to others for building momentum toward sustainability can go a long way toward helping people feel a part of the larger effort and building the political support for sustainability.

Building a healthier social atmosphere and emotional climate to promote sustainability is important to the overall effort. This can be accomplished by using existing committees on campus and opening up discussions about sustainability in general and also by creating a mutually respectful, all-inclusive code of conduct for faculty, staff, and students. In an article in NACUBO's *Business Officer Magazine* on "The Peaceable Workplace," former president of Rollins College Rita Bornstein is quoted as saying that civility is "the art of knowing how to disagree without being disagreeable."[18] The code of conduct increases social health at the institution, teaches interpersonal skills, and reduces, marginalizes, or even eliminates the negative impacts of toxic oppressive personalities who are committed to personal power instead of a sustainable future.

Academic deans and sustainability coordinators, in particular, spend time analyzing the formal and informal power structures at the college and use that knowledge to influence the influencers. When working on passing a proposal, they do more than just hope it will pass. They count their votes and discuss the proposal with key individuals ahead of the meeting.[19] By addressing concerns, gathering insights, and incorporating them into proposals before the meeting, they ensure proposals reflect the group's ideas and will be supported.

2. *Build an effective communication strategy.*

Developing a multipronged communication strategy is another critical step. Through storytelling, peer-to-peer education, and other creative outreach methods, the institutional community experiences multiple messages in a variety of formats. Opportunities to share information verbally and visually throughout the campus include computer monitors in the student center and the recreational facilities, in offices, via the local channel in student housing, and on the campus map and signage. Even bathroom stalls are innovative places to tell both students and staff members the stories of people who are making positive changes, often with small resources. Some universities are developing websites that tell the stories of what students and staff are doing to choose more sustainable lifestyles. The communication campaign can emphasize that sustainability is a personal challenge and national need. Empowered students can help. It was enterprising undergraduates who convinced producers at MTV to create campus contests to "Reduce Your Impact" and "Break the Addiction to Oil."

Still, simply advising first-year students to recycle will not be sufficient. Sustainability awareness does not always equal sustainable behavior. Doug McKenzie-Mohr, author of *Fostering Sustainable Behavior,* describes a critical strategy:

> Community-based social marketing draws heavily on research in social psychology which indicates that initiatives to promote behavior change are often most effective when they are carried out at the community level and involve direct contact with people. The emergence of community-based social marketing over the last several years can be traced to a growing understanding that programs which rely heavily or exclusively on media advertising can be effective in creating public awareness and understanding of issues related to sustainability, but are limited in their ability to foster behavior change.[20]

Community-based social marketing (CBSM) techniques offer specific tools, grounded in research, on how to choose which behaviors to target for change to achieve maximum results, how to investigate the perceived barriers and benefits to change, and how to develop targeted strategies to help foster the desired behavior. A variety of successful models for campus-wide behavioral change programs have emerged including Eco-Reps,[21] sustainability "ambassadors" programs, and laboratory greening programs like the University of California–Santa Barbara's LabRATS.[22] To assist campuses, the U.S. Partnership for Education for Sustainable Development, in collaboration with AASHE and HEASC, has hosted national conference calls with experts to share psychology and marketing research on creating effective environmental behavioral change.[23]

3. *Create systemic changes in the curricula.*

Sustainability needs to be included in all disciplines, programs, and general education requirements. It should not just be confined to an individual major, minor, or degree because this leaves the majority of students illiterate about sustainability challenges and unable to engage in solutions. Students need to be able to think systemically and apply change management skills to create and implement solutions. Faculties also need ongoing professional development resources to build assignments in which students can practice how to help create and implement sustainability actions beyond the classroom. Examples and specific strategies to integrate sustainability into the general education core and into all disciplines are available in the Academic Guidance document published by the American College & University Presidents' Climate Commitment.[24] The Carnegie Foundation for the Advancement of Teaching's book, *Educating Citizens: Preparing America's Undergraduates for Lives of Moral and Civic Responsibility,* documents how our historically integrated and interdisciplinary general education core lost much of its power to teach responsible behaviors and citizenship when it moved to a formula of disjointed choices from a variety of disciplines.[25] A comparison of sustainability learning outcomes in multiple countries showed the international concurrence regarding the importance of learning how to make positive change in the real world.[26]

Through easy entry processes, hesitant faculty can be encouraged to include sustainability in their courses (see table 3.3). After a single orientation session, professors can smoothly combine their traditional course outlines with a sustainability concept and create a learning activity

Table 3.3. Pedagogies and Approaches to Cultivate Lifelong Sustainability Change Agents

- Participatory-action research
- Real-world problem solving
- Interdisciplinary learning
- Problem solving around large societal dilemmas
- Integration of interpersonal skills into group assignments
- Integration of change-management skills practice into assignments

that teaches both. Faculty members are then invited to share their learning activities with each other in a meeting setting. Jean McGregor from Curriculum for the Bioregion created and uses this with great success in her extensive project working with faculty in the Northwest.[27] Debra Rowe has since implemented it effectively at dozens of institutions throughout the country. Hesitant educators can also easily add a positive futuring assignment and have students use their disciplinary concepts to explain how we might grow into a healthier society.[28] Key questions inserted throughout any course, such as "How can we use what we are learning to make a better, more sustainable society?" can help shift the course to a sustainability focus. Still, these approaches need to be combined with a broader strategy to deliver sustainability education to students with rigor to accomplish the desired student learning outcomes. Once faculty members are engaged, they can ask their textbook representatives to include sustainability themes and examples throughout the chapters in the next revision of their class texts. Including sustainability to a greater extent in textbooks from all disciplines will help achieve the national curricular shift that many leadership teams are seeking. The Disciplinary Associations Network for Sustainability has resources that show how this makes good business sense for the publishers.

4. *Institutionalize sustainability.*

It is important to ensure that sustainability moves beyond being an innovation to becoming an established part of a college or university's mission and values. This can be accomplished by building sustainability into the mission and strategic plans annually, by creating structures such as offices of sustainability to manage change and facilitate the shifts required, and by including sustainability in all job descriptions and performance reviews for as many positions as possible on campus. Policies incentivize action, provide a platform for change agents, and encourage those slow to adapt to change to recognize that sustainability is a credible priority for the institution. Institutionalizing sustainability can also be tangibly advanced by using assessment tools such as AASHE's Sustainability Tracking and Rating System (STARS)[29] for a process of continuous improvement. Creative solutions can be found to overcome obstacles. For example, in 2003 students at the University of California–Santa Cruz (UCSC) created a coalition of students, faculty, staff, and farmers called the Food Systems Working Group to bring more local, organic food onto campus. Stalled by complex contract and insurance requirements, stakeholders worked with farmers to form the Monterey Bay Organic Farmer's Consortium (MBOFC). This allowed local organic farmers to meet requirements by operating under the umbrella of ALBA, a worker-supportive operation. The participating growers also agreed to make their farms available for organic farming and food system research conducted at the Center for Agroecology and Sustainable Food Systems at UCSC. Institutionalizing sustainability allows for the ongoing work of identifying barriers to, and creating lasting solutions for, sustainability.

5. *Eliminate systems that are barriers to sustainability, particularly regulatory and financial barriers.*

This key strategy engages students, faculty, and staff members in projects to reduce the barriers to sustainability at the institutional and governmental level. Daniel Kahneman, professor of psychology and public affairs at Princeton University, won a Nobel Prize for uncovering the flaws in the assumption that people optimize their decision making for economic benefits. His work has been integrated into newer economic theory that understands the need to modify our economic regulations for sustainable economic health.[30] Others in the academic community, such as Nobel Prize winner Joseph Stiglitz, the economists in the University of Vermont's Gund Institute for Ecological Economics, and Tim Jackson, author of *Prosperity without Growth: Economics for a Finite Planet,*[31] have led the way in pointing out both the flawed assumptions inherent in our present modification and metrics of the U.S. economy and the opportunities for better policies to build healthier sustainable economies.[32]

However, it is not enough for esteemed scholars to study the need for financial and economic regulatory change. Changes will have to happen at the governmental level to create a sustainable future, and in this process enterprising students will be offered options to investigate present policies and present their findings to lawmakers to help catalyze governmental regulations that can create robust business opportunities for sustainability-oriented products and processes. Presidents and trustees, especially, are grasping new opportunities to use their governmental relations responsibilities to help reframe governmental regulatory and financing systems to make sustainability easier to accomplish. Examples of these initiatives include the following—integrating student participation where possible:

—An energy policy to stop the uneven subsidization of fossil fuels over renewable energies and efficiency

—A move from counterproductive regulations in the funding of governmental building to policies that support cost-effective green building for all new construction and renovations

—Local watersheds analysis and presentation of the analysis and related policy recommendations to city and regional councils

—Media relations that distribute higher education sustainability policy examples efficiently to the local community for replication, including feeding the information to national news sources so it can serve as a model for the nation

—An engaged community visioning and implementation process for sustainability that systematically includes key government, business, and non-profit stakeholders.[33]

To make systemic change, it is important that what happens on campus not *stay* on campus. Working with the community magnifies the changes that can be made. *An institution-wide initiative to engage in real-world problem solving for sustainability helps foster this engagement with the larger society.*

Conclusion: The Skills and Expertise Sustainability Leaders Now Need to Develop

Effective change agents are persistent and resilient in assessing the evolving needs of the institution and working to implement transformational programs. Sustainability leaders build strong coalitions. They inspire campus members at all levels of the academic and administrative hierarchies and in all constituencies, from students to trustees, to become change agents themselves. They benefit greatly from solid grounding in organizational and behavioral change literature and familiarity with emerging research in how to foster sustainable behavior in people's actions regarding social, economic, and environmental responsibility. Effective sustainability leaders make the effort to think and plan strategically to move systems at a deep level while remaining flexible enough to incorporate new information as sustainability itself evolves in and around higher education. They take the time to catalyze both formal and informal power structures to support sustainability, and they are able to manage multiple campus efforts simultaneously and keep moving forward despite setbacks, all to preserve an institutional trend of increasing momentum toward sustainable thinking and action.

As outlined by the ACPA's Presidential Task Force on Sustainability, the most effective change agents are "resilient, optimistic, tenacious, committed, passionate, patient, emotionally intelligent, assertive, persuasive, empathetic, authentic, ethical, self-aware, competent, [and] curious. . . . Change agents help envision, articulate and create positive scenarios for the future of society. [They] see the paths, small steps, for changes needed for a more sustainable future, convert it into a task list and timeline, and follow through effectively."[34] Particularly when the change agent is a sitting president, that individual can be most effective when he or she facilitates the commitment in community members to develop their own visions of sustainability. In this regard, presidents need to be leaders who are skilled negotiators, facilitators, and mediators.

Whether as president, dean, or sustainability director, change agents use optimism to create more successes. Often the leader will initially establish expected desired principles of conduct. These can include creating atmospheres of mutual respect, the expectation of continuous problem solving, and simply active listening. In developing their cultures of sustainability, some colleges are incorporating positive psychology and a commitment to building interpersonal communications skills into their plans for institutional transformation over the coming decade. Academic deans and faculty leaders in various disciplines, especially communication, psychology, and organizational behavior, can contribute to sustainability goals and objectives by developing syllabi that build skills for interpersonal and organizational health. One example of an international resource useful in this kind of course and program planning is the Center for Nonviolent Communication (CNVC). Numerous training opportunities are offered in communities around the globe. In the words of founder Marshall Rosenberg, "NVC guides us in reframing how we express ourselves and hear others. Instead of being habitual, automatic reactions . . . our words become conscious responses based firmly on an awareness of what we are perceiving, feeling, and wanting. We are led to express ourselves with honesty and clarity, while simultaneously paying others a respectful and empathic attention. . . . The form is simple, yet powerfully transformative."[35]

Effective sustainability leaders understand their responsibility to utilize the media and public relations resources to inform the public and help them create sustainable communities beyond their campuses. Leading by example through personal and professional decision making is also a resource that is more meaningful to colleagues than some may realize. Debra Rowe told attendees at the Kansas State statewide sustainability conference in 2010:

> Love yourself as passionately as you do sustainability. Take moments to revitalize. Do not allow yourself to burn out. Sculpt an internal voice of self-encouragement and nurturing, self-acceptance, and forgiveness for what you are still struggling to accomplish. Take moments when the feelings build up to grieve the bigger picture of lost sustainability opportunities, the ecosystem degradation and the human suffering it has caused, but don't let the

sadness overwhelm you. Celebrate your efforts and what you are trying to accomplish, both in public and in private as appropriate.[36]

In sum, leadership for sustainability is based on positive possibilities. When infants eighteen months old see an unrelated adult whose hands are full and who needs assistance opening a door or picking up a dropped clothespin, they will immediately try to help.[37] As Frances Moore Lappé explains, "We are hardwired for cooperation."[38]

We have only recently begun, in higher education, to understand deeply the importance and value of building healthy self-concepts and interpersonal skills as part of our education.[39] Much of the challenge of sustainability leadership is to develop cultural, political, economic, and educational systems that cultivate the best in us. Growth in higher education's teaching and learning of sustainability and related changes in administrative structures can create fertile soil for local, regional, and national policies, corporate practices, and behavioral shifts that are rooted in sustainability thinking.

As higher education moves forward to educate for a sustainable future, it is important that we do not wait for perfect solutions. Each of us has opportunities now to stimulate sustainability action at multiple levels of our university or college. Each institution needs to move forward with a customized and multipronged approached that is adapted to its own challenges and opportunities while remaining connected to any larger systemic shifts. This chapter has provided a set of resources, strategies, and perspectives to help presidents, deans, and all campus sustainability leaders move forward with both small and large steps for success.

CHAPTER FOUR

Measuring Campus Sustainability Performance: Implementing the First Sustainability Tracking, Assessment, and Rating System (STARS)

Judy Walton and Laura Matson

The sustainability movement in higher education has seen explosive growth since 2000. Seeds planted in the 1990s have taken root and burst into a healthy network of organizations, programs, and campus initiatives. One sign of the movement's evolution is a growing interest on the part of senior campus officers such as presidents, provosts, and chief financial officers to assess the effectiveness of sustainability initiatives across institutions. In the summer of 2006, the Higher Education Associations Sustainability Consortium (HEASC), a relatively new network of higher education associations, issued a call for the development of a standardized campus sustainability assessment tool. The call came in response to a growing number of colleges and universities claiming to be sustainability leaders as well as the rise of third-party rankings and top ten "green campus" lists. While an array of individual or regional campus sustainability assessment tools already existed, they varied widely, and there was no agreed-upon standard.

The HEASC associations envisioned a comprehensive system that would "address all the dimensions of sustainability (health, social, economic and ecological) and all the sectors and functions of campus, including curriculum, facilities, operations, and collaboration with communities." They wanted a tool that any higher education institution could use. With that in mind, HEASC called upon the Association for the Advancement of Sustainability in Higher Education (AASHE) to "convene all relevant stakeholders in a collaborative process to develop such a system."[1] The result, developed through a three-year collaborative process led by AASHE, is the Sustainability Tracking, Assessment & Rating System (STARS), a voluntary, self-reporting system that provides recognition to participating institutions and enables them to benchmark

Three Emerging Trends

- Incorporate sustainability throughout the entire college curriculum.
- Leverage student talent in meaningful ways, including participation in key decisions and policies.
- Expand research opportunities as new streams of government funding support sustainability research across disciplines, including behavior change, resource policies, and new technologies.

their progress over time as well as compare to other institutions. Participation is open to all colleges and universities in the United States or Canada. STARS credits are designed to engage and reward institutions at any stage of sustainability, whether long experienced or just getting started.

Initial Goals of STARS

STARS represents an innovative step forward in higher education assessment in a number of ways. Its iterative and highly participative process of development included reviewers from nearly every sector of higher education. This process, along with its commitment to transparency and its rigorous insistence on integrity and adaptability of the system, make it one of the most progressive assessment tools in higher education. Few other tools require such a high degree of public transparency and accountability at every level, including sign-off by the president or chancellor. STARS is also the first self-conducted assessment system to enable meaningful comparisons across institutions using agreed-upon metrics. In addition, STARS offers these four advances:

1. A *framework,* or road map for understanding sustainability in all sectors of higher education, from curriculum and research to operations, investment, and community engagement

2. A means of *benchmarking* sustainability progress at individual institutions over time

3. Compelling *incentives* for continual improvement toward sustainability, with a point system that recognizes even small degrees of progress and a means of peer comparison that motivates institutions to move forward through friendly competition and collaboration

Three Best Practices

- Perform regular benchmarking and assessment to set priorities.
- Celebrate accomplishments.
- Involve stakeholders across campus so that there is a shared culture of sustainability.

4. A robust means of *information-sharing* so that every college or university can learn from the sustainability practices and performance of reporting institutions.

Finally, STARS aims to build a stronger, more diverse national campus sustainability community through its comprehensive coverage of all operational sectors and its commitment to include the broadest dimensions of sustainability, from environmental health to social justice.

To help colleges and universities reap these benefits, support from each participating institution's leadership is critical. Buy-in from presidents, provosts, and even board members helps ensure that disparate campus sectors that are important to STARS but do not typically collaborate (e.g., purchasing, human resources, and waste management) come prepared to engage with and learn from each other. Oversight by the members of the president's cabinet is also important, given the requirement for public disclosure of information and the need for accuracy.

More specifically, direct involvement by the chief executive officer or provost is required at two steps during the STARS participation process. First, upon registering for STARS, the applicant must provide the name and email address of an executive at the institution, typically the president or chief academic officer. This individual receives a registration confirmation message informing him or her of the institution's participation. Second, upon submitting for a STARS rating, the institution's chief academic officer must provide a cover letter that vouches for the accuracy of the information included in the STARS report. Providing this letter is the final step in the STARS reporting process, and once the letter is received the reported data are made available publicly on AASHE's website.

Starting the Program

For a year and a half after the call by HEASC described above, AASHE gathered feedback from participants in workshops held at multiple campus sustainability events. This information provided the

Single Most Important Piece of Advice for a New Professional

- Do not be afraid to reach out for help. Learn from your peers at other institutions and seek out available resources. Move quickly but without compromising quality.

groundwork for the first publicly available draft version, STARS 0.4, released in September 2007. Experts from a cross-section of the sustainability and higher education communities provided useful comments on STARS 0.4 through surveys, focused conference calls, and submissions of opinions. All of this information was used to develop a pilot version of STARS, which was released in two phases in 2008. Nearly seventy universities and colleges completed the year-long pilot project, testing the system and providing extensive additional comments and enhancements. Those summary results were posted on the STARS website. While the pilot project was underway, AASHE released the next draft version for the public, STARS 0.5, which again generated thoughtful responses and suggestions from diverse higher education and sustainability stakeholders.

In September 2009, AASHE launched STARS 1.0 Early Release, which incorporated constructive comments, improvements, and lessons learned from the three previous years. While STARS may very well be the most thoroughly vetted and extensively tested campus sustainability framework in North America, it intentionally remains a work in progress, as stated in the Technical Manual: "STARS 1.0 is intended to stimulate, not end, the conversation about how to measure and benchmark sustainability in higher education."[2] AASHE has continued to revise the system over time to ensure that it remains as useful and effective as possible for new cohorts of institutions joining the program.

Toward that end, AASHE developed a governance structure to ensure that STARS responds to the evolving and future needs of participating colleges and universities. AASHE's Board of Directors—composed of college and university administrators, faculty, staff, students, and business leaders—serves as the ultimate decision-making body on matters related to program policy. In addition, the STARS Steering Committee, made up of campus practitioners and several members of AASHE's Board of Directors, advises on program policy and serves as the decision-making body on matters related to the content of STARS credits. For example,

A Decision One Would Make Differently with Hindsight

- Aim for a better balance between work and play.

when AASHE desires to update credits to reflect new developments in the field, the Steering Committee decides when and how each credit should be updated. The third part of the governance structure is a Technical Advisors Group, which includes individuals with subject matter expertise appointed by the Steering Committee, who provide official recommendations on interpretations of and revisions to the STARS credits. The individuals who serve as members of these three groups are drawn from varied institution types, locations, and departments to reflect the diversity of higher education sustainability stakeholders.

STARS and the Concept of Sustainability

The concept of "sustainability" has shaped the development of STARS and is fundamental to the mission and goals of the rating system. While sustainability has become an increasingly popular term both on campus and in society, its meaning and history have not always been clearly understood. AASHE uses the definition that is most common and inclusive, encompassing *human and ecological health, social justice, secure livelihoods, and a better world for all generations.* This definition draws from what is perhaps the most widely cited origin of the term: an April 1987 report from the UN World Commission on Environment and Development called *Our Common Future,* which is often called "the Brundtland Report" in recognition of Gro Harlem Brundtland, then commission chair and Prime Minister of Norway. While the report never defines the word *sustainability,* it does define "sustainable development" as "development that meets the needs of the present without compromising the ability of future generations to meet their own needs."[3]

Throughout *Our Common Future,* the interconnectedness and interdependence of the social, environmental, and economic components of sustainability are emphasized, as is the relative neglect of economic and social justice. Most definitions of sustainability today continue to emphasize its simultaneous consideration of economic, environmental, and social dimensions. Corporations, for example, focus on the "triple bottom line"—a simultaneous consideration of "people, planet, and profits" (or "human, natural, and financial capital"). This understand-

A Sustainability Myth

- Sustainability costs too much or "our campus can't afford sustainability."

ing is also reflected in sustainability rating systems for business such as the Dow Jones Sustainability Index.[4] STARS represents an attempt to translate this broad and inclusive view of sustainability into a comprehensive set of measurable objectives at the campus level. This is why STARS credits encompass an institution's environmental, social, and economic performance.

Details of the STARS Framework

Establishing Rating Levels and Timeframe

There are four STARS rating levels: *Bronze, Silver, Gold,* and *Platinum.* In addition, a Reporter designation is available for institutions that do not wish to pursue a rating but do wish to use the reporting system as a repository for their sustainability data and earn recognition for publicly reporting. Only positive ratings are available through STARS; each rating level represents a significant achievement worthy of recognition and shows sustainability leadership. Participating in STARS involves gathering extensive data and sharing it publicly—a commitment to sustainability that should be applauded.

STARS is designed to incorporate all levels of sustainability achievement, with the highest levels of achievement representing ambitious, long-term goals for any institution no matter its resources and history. It is expected, therefore, that in the future only a few institutions will achieve certain credits. A STARS rating is good for three years, but institutions may update information in their profiles and pursue new ratings as often as once a year. Additionally, it is expected that institutions will make a good faith effort for the duration of the STARS rating to maintain the status that made them eligible for a credit.

Selecting and Vetting Credits

The initial set of STARS credits was developed in large part through AASHE's review of campus sustainability assessments, sustainability reports from businesses, and various sustainability rating and ranking systems. Based on comments from several hundred experts and stakeholders in several draft rounds, the initial credits were revised over time or eliminated and new credits were added until the first official version, 1.0, was ready to launch. To ensure that the system would work as intended, AASHE strived to make the credits *objective, measurable,* and *actionable.* In addition, all credits were vetted using three criteria. First, in order to be included, each credit had to lead to improved environmental, social, and/or economic performance by colleges and universities. While the place of impact may vary—for example, institutions can accelerate the transition to renewable energy systems by installing technologies on campus, investing endowment funds in renewable energy companies, teaching students about renewable energy, or conducting research on new technologies—each credit should indicate a movement toward sustainability.

Second, each STARS credit had to be appropriate for the vast majority of institutions despite the diversity of institutional types and circumstances. In order to accommodate this diversity, some credits do not include detailed specifications but are instead flexible and open. Others include an applicability criterion, which means the credit only applies to certain types of institutions. By following this approach, institutions are not penalized when they do not earn credits they could not possibly earn due to their circumstances. For example, a credit related to activities in residence halls would not apply to community colleges that do not maintain residence halls.

Third, performance was prioritized over strategy wherever possible in STARS. Performance credits are based on measurable sustainability outcomes. These are often, but not always, quantitative, such as the percentage of employees who use alternative modes of transportation to travel to and from campus. Strategy credits, conversely, focus on approaches or processes that may help improve an institution's performance or outcomes, such as offering transit passes or operating a campus shuttle service. A university may take different strategies or approaches to achieve the same performance outcome. While both types of credits provide useful information, the primary goal of the system is to catalyze tangible improvements or outcomes, not simply to encourage adoption of strategies that may or may not achieve the desired outcomes. In instances where measurable, meaningful, and comparable performance indicators could not be identified, a strategy credit was often used instead. In some ways, strategy indicators can tell a richer story about an institution's sustainability initiatives and provide valuable information that is worth collecting and sharing. As a result, both strategy and performance credits have a place in the system.

Weighting the Credits

After the set of credits was firmly established, each credit was weighted relative to others and assigned a point value. Points were allocated using the following considerations:

—To what extent does the credit contribute to improved environmental impacts?

—To what extent does the credit contribute to improved financial impacts?

—To what extent does the credit contribute to improved social impacts?

—To what extent are there educational benefits associated with the achievement of this credit?

—Breadth: How many people are affected?

—Depth: How deeply are people affected?

As these questions indicate, the focus in allocating points was on the *impact,* not the *difficulty,* of earning the credit. Some sustainability initiatives may be very difficult to implement but yield negligible effects. Conversely, some generally easier projects can have significant effects. Assigning points based on the difficulty of earning a credit would create a perverse incentive for institutions to focus on the difficult projects or initiatives that may not necessarily be the most meaningful or have the most effect.

While AASHE strived for a fair and consistent approach to allocating points, such an exercise is inherently subjective. Developing a more robust point allocation methodology, including finding stronger ways to account for the influence of regional variations and differences in institution type, will be considered for future versions of STARS.

Collecting and Reporting Data

All reporting for STARS happens online through the STARS Reporting Tool. No hard copies of materials or paperwork are required. Institutions are expected to include their entire main campus when collecting data and may also choose to include other land holdings, facilities, farms, and satellite campuses as long as the selected boundary is the same for each credit. The boundary is specified along with the credit data in the online STARS Reporting Tool. If an institution finds it necessary to exclude a particular unit from its data, it must provide a reason for doing so in the notes accompanying the submitted data.

As part of a strategy to ensure accuracy and accountability, a "responsible party" must attest to the accuracy of each STARS credit during the reporting process. In order to sign off on the credit in the Reporting Tool, the institution must provide the contact information for a responsible party who affirms the accuracy of the data included in the submission. Institutions have one year from the date they register for STARS to complete the reporting process and receive a rating.

Ensuring Accountability

Although STARS is a self-conducted assessment, in the future AASHE may pursue opportunities for third-party verification. Initially, STARS incorporates three strategies to ensure that submitted information is accurate:

1. Most information and documentation that institutions submit to achieve a rating are made publicly available through the AASHE website.
2. For each credit, an attestation is required from a responsible party affirming the accuracy of information submitted through the Reporting Tool.
3. Each submittal for a STARS rating must be accompanied by a letter from the president or chancellor that affirms the accuracy of the institution's report. Sign-off from the institution's chief executive is an important strategy for promoting accuracy within the STARS self-assessment framework.

The president's letter also serves as an introduction or cover letter, and may include a description of the college or university's commitment to sustainability, background about the institution, key achievements or highlights from the STARS report, and goals for future STARS reports.

Why Participate in STARS?

From recruitment and planning to community learning and institutional efficiency, participating in STARS can help institutions meet diverse goals.

Improved Image and Enhanced Recruitment

STARS helps colleges and universities communicate their sustainability accomplishments and demonstrate sustainability leadership to alumni, local communities, and prospective students. This leadership can help boost student recruitment. Prospective students are increasingly indicating that sustainability is an important criterion in selecting the institutions to which they apply and attend. The Princeton Review's 2008–09 College Hopes & Worries Survey asked, "If you (your child) had a way to compare colleges based on their commitment to environmental issues (from academic offerings to practices concerning energy use, recycling, etc.), how much would this contribute to your (your child's) decision to apply to or attend a school?" Two-thirds of respondents affirmed interest in having that information, and 24 percent said it would "strongly" or "very much" influence application and enrollment decisions.[5] Similarly, a survey of freshman attitudes conducted by the Higher Education Research Institute at the University of California, Los Angeles, found that nearly half (45.3%) of the respondents believe "adopting 'green' practices to protect the environment" is "essential" or "very important."[6]

Guidance for Planning and Visioning

STARS can also help institutions develop sustainability plans and set related goals. Since the system encompasses long-term sustainability goals ranging from carbon neutrality and zero waste to offering robust and diverse sustainability courses and research activities, STARS can serve as a tool to help campus members envision what a sustainable university or college looks like and what it will take to move toward that goal. Since STARS presents an organizational framework for understanding the dimensions and diverse aspects of campus sustainability, the system is also useful for institutions starting sustainability offices or committees by providing a template for how to organize these efforts.

Efficiency

For years, many institutions have conducted independent and internal sustainability audits or assessments in an attempt to benchmark and set goals. While participating in STARS is not a simple undertaking, it may be more efficient than conducting an independent assessment since the indicators have already been selected, guidance on collecting and tracking information is provided, and an online reporting system is available that can serve as a repository for data and an organizing tool.

Benchmarking against Peers

The ability to set goals and develop plans for sustainability is enhanced by the capacity for peer-to-peer benchmarking that STARS fosters. By using a common standard of measurement and having submissions be publicly available, participating in STARS can help institutions identify best practices among peer institutions, share lessons learned, and stimulate a little friendly competition. This can help institutions identify strengths and areas for improvement and allow them to communicate their sustainability achievements publicly.

Enhanced Organizational Learning

STARS can also serve as a tool for creating a campus learning community more focused on sustainability thinking. Since the system incorporates multiple dimensions of campus sustainability that touch disparate campus sectors, STARS brings many stakeholders together under a common framework. Specifically, working on the STARS assessment gives the institution's STARS Liaison an opportunity and reason to connect with departments and learn about the sustainability activities already underway. In this way, campus leaders can use the tool to help break down departmental silos and promote the cross-disciplinary thinking and work that is integral to sustainability. Furthermore, since an institution's online STARS report is made public, it can serve as a central resource for all campus community members interested in learning about the college or university's activities and in identifying opportunities for their departments to contribute to broader sustainability initiatives.

As part of his master's degree work, Kyle C. Murphy coordinated the Evergreen State College's participation in the STARS Pilot Project. In his thesis, Murphy explored whether Evergreen's use of STARS led to organizational learning. He found that the project provided new information and involved multiple stakeholders in decision making, two conditions that support effective, long-term organizational learning. As evidence that organizational learning occurred, Murphy noted that participation helped streamline processes for tracking the same information and that community members developed new understandings of campus sustainability efforts and identified new opportunities for collaboration.[7]

Enhanced Student Learning

In addition to organizational learning, STARS can serve as an effective tool for student learning both in and beyond the formal curriculum. Reports indicate that students contributed to their institution's participation in the STARS pilot project in multiple ways. Some students, like those at Evergreen, coordinated their institution's participation in the pilot project as part of a master's thesis or senior capstone project. Others, including some community college and graduate students, coordinated their institution's participation in the pilot project through an internship or employment with the sustainability office. A large number of institutions enlisted student interns and staff to assist with data gathering and compilation.

Opportunity to Shape the System

Since STARS is an ongoing project and will be continually updated in response to user feedback, participating in STARS enables schools to help shape the overall framework nationally. There are multiple opportunities for engagement with the system, and directly participating means that an institution will be able to help determine what is included and how it is counted in future versions.

Timeframe for Institutional Participation

The amount of time it will take to complete STARS will vary from institution to institution depending on several factors:

—How much information the institution is already tracking (e.g., has the institution completed a greenhouse gas emissions inventory?)

—Which credits the institution is attempting (institutions are required to submit data only for the credits they are pursuing)

—How centralized the institution is (e.g., will data gatherers have to approach many departments to get information about paper purchasing, or is paper purchased centrally?)

—Whether this is the first time participating in STARS—getting acquainted with the logistics of the system and the details of the credits and process may require some extra time during the first assessment.

Participants have one year following registration to complete the submission, although institutions may request a six-month extension. Building enough time into the schedule to allow data-holders to assemble and approve the information is important. Likewise, since the final step in the process is to post a letter from the president or chancellor affirming the veracity of all information submitted, the time required for this step should be accounted for as well.

To help make STARS an effective tool for fostering community conversation, institutions are encouraged to share results from the assessment with the relevant stakeholders and the campus community broadly. After completing the assessment, institutions may wish to present the results in a public forum on campus or to the sustainability committee. This time can be used to explore the results and set goals for the future.

Some Lessons Learned by Early Participants

Before beginning a STARS assessment, it is helpful to generate support from various stakeholders on campus. Since the assessment covers multiple topics and the activities of various departments, an institution's participation will be more successful if there is high-level support or endorsement and if enthusiasm for participating is shared across campus, including faculty, administrators, staff—and, importantly, members of the board where feasible.

Once ready to begin, the first step is to identify the official STARS Liaison, who will serve as the primary communication contact with AASHE on all STARS matters. Although each institution is required to designate a Liaison, the approach taken to gathering, assembling, and verifying data included in the STARS report can vary across campuses. Some institutions enlist the sustainability committee to tackle the data collection process, while others engage student interns or students enrolled in a sustainability course. The Liaison simply communicates with AASHE as needed and disseminates information from AASHE to the data collectors. When setting up a system for gathering and reporting data, it is important to stay organized and establish good procedures to keep track of all the incoming information; the online Reporting Tool has been designed to help with this task. Involving students in the STARS data collection process can serve as an excellent learning opportunity. When working with students as data gatherers, it is helpful to set clear expectations and to develop a memorandum of understanding clarifying how the data can and cannot be used.

After assembling the STARS team on campus, the next step is to identify who to contact for the various data. At some universities, much information may already be tracked centrally within the sustainability office or the department of institutional research. At most institutions, however, several different departments will be called upon to contribute information. As such, it is helpful for administrators to notify the affected departments directly and ahead of time to let them know about the benefits of participating so they can be prepared to supply information. In addition to notifying campus stakeholders of the institution's sustainability commitment and leadership, this advance message can also improve response and participation rates from the many campus data holders. The STARS coordinator or liaison may wish to connect with the institutional research office to determine what data are already being tracked centrally and what other departments hold relevant information.

Conclusion: The End of the Beginning

It is still quite early to assess the full impact of STARS in advancing sustainability on college and university campuses. Already, however, several effects of the program can be felt. A number of colleges are viewing the data collection system as a solution to survey overload caused by the proliferation of third-party "green campus" rankings and their accompanying questionnaires. Conversations are also underway to use STARS as a central repository where campuses may direct third-party data gatherers and eliminate the need to fill out what are often long and detailed surveys. Furthermore, some universities have downloaded the STARS Technical Manual and are using it as a guideline to perform a campus assessment without registering to use the Reporting Tool and submitting their information formally.

In online discussions, conferences, higher education media, and journal articles, the conversation is turning toward STARS and its credits. Over a dozen higher education associations and organizations have signed on to become "STARS Partners" and to promote use of the system among their member institutions and provide training and resources in their areas of expertise. In this sense, STARS is already moving toward its goal of serving as a "framework or road map for understanding sustainability in all sectors of higher education." With time, it is anticipated that the system will also fulfill its goal of "facilitat[ing] information sharing about higher education sustainability practices and performance" so that all colleges and universities can learn from the sustainability achievements of STARS participants.[8]

While the ultimate success of STARS in meeting its goals is still to be determined, it is clear that the system is already helping many campuses set and meet sustainability goals while helping to shape the conversation about sustainability in higher education. Moreover, given the system's intent to evolve over time and improve in response to the needs of the higher education community, it is likely that STARS will serve as the standard to assess sustainability progress for this generation of leadership.

CHAPTER FIVE

Institutionalizing Sustainability: Achieving Transformations from the Inside

Leith Sharp and Cindy Pollock Shea

Implementing the Vision

An effective sustainability vision must not only address direction and purpose but, perhaps even more importantly, it must address the means by which colleges and universities will achieve real and lasting progress. That is to say, an effective sustainability vision must address not only the *what* and the *why* but also the *how.* A good number of higher education institutions still define their sustainability vision only in terms of a set of environmental impact reductions goals such as greenhouse gas reductions, waste reduction goals, and green building goals. They are strong on the *what* and the *why,* but they simply ignore the *how.* This tendency involves a degree of wishful thinking that the *how* will simply involve tinkering at the edges of a college or university's practices.

In reality, simply tinkering is not an option if we want real and lasting progress toward sustainability. We must instead pursue a variety of internal reforms to remove a myriad of institutional barriers to innovation and good business practices. Without them, we risk having all of our leadership, funding, and effort directed immediately toward the quick fixes that some technology offers us without building a foundation of greater organizational capacity. Many universities and colleges have invested funds into upgrading lights, building controls, and other technologies without involving and training the building managers or engaging the faculty and staff in changing their behaviors to complement these new systems. The result is often money spent without any real financial or environmental return. Another common story involves facilities staff who know of numerous potential energy conservation projects on campus and yet are unable to access any funds regardless of

Signs of Sustainability Success on Campus

- Employees are hired, trained, and evaluated based at least in part on their ability to advance sustainability.
- Students graduate with an understanding of sustainability principles and how to apply them.
- Members of a given university, college, or community college know, understand, and work to promote the sustainability goals of the institution.
- Innovation, collaboration, and managed risk are valued within the campus culture.
- Knowledge and new ideas are encouraged and appreciated and move fluidly up and down the organization.
- Effective change management techniques are applied widely and often to foster and reward continuous improvement.
- Purchasers consider the economic, environmental, and social impacts of their decisions.
- Food is produced locally, using sustainable methods.
- Buildings are managed to optimize energy efficiency as well as provide occupant comfort.
- Non-potable water is used for irrigation, toilet flushing, and fire suppression.
- At new campuses, wastewater is treated on site.
- New buildings are powered by renewable energy.

the extraordinary payback opportunities. In these instances, finance and accounting systems work against good business practice.

As one institutional best practice, leadership teams can directly call for campus-wide engagement, new modes of continuous improvement, a culture of practice-based learning, and a spirit of collaboration. Beyond this, presidents should also put in place new financial incentives, goals and reporting requirements, and rewards and recognition opportunities. In doing this, they will ensure that some of the early leadership, funding, and effort is properly invested in solving the *how* as well as the *what.* In addition, the trustees and faculty should both review and renew any vision statement every three to five years and invite representatives from across the institution to comment on what the new vision should be for the next five years or more.

Sustainability and Effective Strategic Planning

Given the complexity of campus sustainability initiatives today, it is likely that a variety of strategic planning efforts will be necessary, focusing on multiple operational and impact areas. Greenhouse gas reduction plans have become one of the most common campus sustainability strategic plans; however, a waste reduction plan, a green building plan, and a water conservation plan will also be necessary at some point in the campus sustainability evolution. In addition, an overarching strategic plan for driving campus-wide sustainability will be necessary, along with assessments of its outcomes and effectiveness. Strategic planning efforts become important vehicles for engaging a diversity of campus stakeholders, including board members, senior administrators, students, faculty, and alumni in a productive, collective process. Strategic planning for campus sustainability should become an ongoing activity. All plans should be continuously revised and strategic planning documents should be considered to be living texts.

In reality, colleges and universities have an unconscious life, just as individuals do. They have established habits that have long gone unquestioned. In certain arenas they can exhibit irrational tendencies and inconsistent behavior. Submerged just below the surface can be deep internal contradictions in purpose. Complexity, time limitation, personalities, egos, power struggles, and differing assumptions regarding how things work can all act at different times to fuel the flames of organizational dysfunction. Strategic plans are important instruments for setting forward a pathway for progress; however, they can assume a higher degree of rationality and transparency than actually exists in many higher education institutions. Common risks that undermine the effectiveness of strategic planning for campus sustainability include the following:

—The scope of the planning effort is too broadly defined.

—The time span for planning, assembling, and writing is too short.

—The right people are not engaged effectively.

—The planning process is not facilitated effectively and there is not enough listening, collaboration, and tension tolerance.

—Competing agendas and lack of goal clarity drive participants into conflict and unnecessary conservatism.

Some colleges begin by using pilot programs to test new technologies and design and to diagnose organizational implications and requirements for larger-scale success. Pilot programs can be effective in resolving conflicting assumptions, revealing capacity-building needs, ensuring accuracy regarding financial implications, and determining leadership expectations.

New Models of Institutional Leadership

In particular, presidents and provosts need to understand that the complexity of addressing sustainability at institutions of any size and history demands a new model of leadership. In order to make real progress toward sustainability while maintaining organizational equilibrium and vitality, leaders must develop new skills, capacities, and processes to define and solve problems that reach beyond traditional boundaries and management areas. Solutions must emerge from effective processes of collaboration, systems thinking, and practice-based continuous improvement to ensure stability. In addition, campus leaders must grasp deeply the importance of valuing the earth's limitations as a compass for guiding the organization's path forward. Just as the many societal systems such as regulation, economics, and culture provide real limits and opportunities for the organization, the earth's life support systems must be viewed as providing dynamic limits and opportunities for the college or university.

Presidents and chief academic officers, especially, will need to set a course for engaging the right participants in processes that are effective in navigating the organization forward. The culture of the organization must then prioritize trust and participation in order to draw the best from all participants. One of the most important leadership roles in accomplishing sustainability objectives will occur in fostering and sustaining a context of collaboration. The complex, systemic, and urgent nature of our many sustainability problems requires a new level of team functioning, multidisciplinary thinking, interdependence, and cross-departmental integration. Without strong leadership driving collaboration, the predominant habits will most likely prevail. As David Chrislip observed in *The Collaborative Leadership Fieldbook,* "Collaboration needs a different kind of leadership; it needs leaders who can safeguard the process, facilitate interaction, and patiently deal with high levels of frustration."[1]

Even on campuses, leadership has long been viewed as coming from the top down in a linear chain of command. In reality, many leadership pathways are more cyclical, involving a leadership system that includes different kinds of leadership coming up from the grass roots and moving horizontally across peer groups. Leadership through the use of authority and leadership through the use of influence are both very important sources of leadership for organizations. Yet a linear perspective of leadership must be replaced with a new understanding of the real ways in which leadership works across and between all tiers of the university or college for sustainability thinking to succeed over the long term. With the emergence of these new leadership competencies, priorities and perspectives will open the way for new governance structures to emerge that are much more effective.

Better Governance for Campus Sustainability

Governance for campus sustainability must be integrated into both the traditional hierarchy and a variety of supplementary governance mechanisms. Most universities and colleges start by establishing a sustainability committee with broad representation from students, staff, and faculty. This can be a helpful start, but it will not be enough to achieve more than marginal progress. The journey toward effective campus sustainability policies and practices requires a deep level of integration into a large number of existing governance mechanisms, including the capital approval process for all buildings, the annual budget processes of all departments, and the ongoing purchase contract management processes of procurement staff members. This integration involves the development of environmental goals, related strategic plans, accountability, and reporting requirements. Only then can the integration be driven from the executive through middle management and down to the grass roots. Effective modes integrate the chain of command along with certain additional new mechanisms like governing committees. A common example is the emergence of a new greenhouse gas reduction goal and the establishment of an oversight committee to review progress and make high-level recommendations.

The institutional sustainability agenda can only become integrated into traditional governance structures of the organization if there is a corresponding expansion of communication pathways for feedback to travel up, down, and across chains of command. Beyond ensuring integration into existing governance structures, it is also essential to support the proliferation of new working groups tasked with advising, energizing, and driving specific programs and projects. Working groups often emerge for green athletics facilities, laboratories, procurement, and the endowment, for example. The skillful facilitation of these groups can ensure the effective engagement of a wide variety of important stakeholders in the process of continuous improvement toward the institution's broad sustainability objectives. Eventually, as the senior leadership develops a greater number of formally stated environmental goals, they are taken up by additional governance groups to help organize and focus their own efforts. Ideally, these grassroots groups become important advisory assets to the presidents and deans as they consider future goals.

The Impact of Institutional Culture

College and university cultures can range from cultures of mistrust and deeply entrenched silos to those of trust, openness, and collaboration. Some campuses become extremely risk averse, while others are risk tolerant. Like many other aspects of higher education organizational life, culture has largely gone underground. In the case of sustainability thinking and action, the work of bringing them back into the light of day is best done through piloting various new projects as a controlled exercise in challenging various cultural traits, forcing them to be articulated, and allowing them to be gently challenged and changed on a small scale through positive experiences of innovation.

Some of the most important campus cultural shifts that leadership teams must achieve to put in place durable sustainability programs and policies include:

—A culture of collaboration, trust, and inclusion

—A culture that is risk tolerant and that values the process of continuous improvement

— Greater distribution of leadership across institutional groups to stimulate wider responsibility

—A culture that is tension tolerant and comfortable with greater degrees of uncertainty and conflict

—A culture that invites different perspectives and seeks to understand the different contexts in which others operate.

Keys to Successful Funding Approaches

One of the central leadership challenges for presidents, chief finance officers, and board members will be financing campus sustainability initiatives over multiyear periods. Up until the mid-2000s, campus sustainability was viewed as another issue competing for precious financial resources that would be better spent on teaching and research. More recently, this assumption has been directly challenged by a large number of green campus programs and projects that have not only paid for themselves over a short period of time but have also produced significant savings over their lifetime. Thus, as of 2010, it is possible to view campus sustainability as an investment rather than a cost. This investment will eventually pay itself back not only financially but in a variety of other ways, such as enhanced student recruitment, staff retention, improved community and neighborhood relations, increased donor support, enhanced institutional image, and regulatory benefits. Perhaps the most common mistake made by institutional leaders is that of underinvesting in campus sustainability. Underinvestment will ensure under-achievement and lack of an adequate return.

There are two aspects of campus sustainability that require financing. The first is the investment in the change management function that will act as the engine driving continuous improvement. This is a human resource investment to be maintained at least for the life of the sustainability effort. Almost all campuses typically underinvest in their change management teams and as a result fail to achieve the kind of pace and volume of campus-wide innovation and cost effectiveness that would otherwise be possible. The second aspect of campus sustainability that requires funding involves new projects and programs that will stimulate new technologies, designs, and institutional practices. Typically, sustainability staff members find it very difficult and time consuming to access seed funding to conduct research and pilot projects to test new practices and chart a course toward greater efficiency and lower environmental impacts. As a result, higher education institutions have engaged in relatively little innovation.

Accounting structures often separate responsibilities for large capital investments from annual operating budgets, incentivizing capital budget managers to pay no attention to potential operating savings and locking operating budget managers out of capital investment decision-making processes. Organizations rarely factor in the operating costs associated with decisions in the large or small arenas of financial planning, and the focus on front cost only makes them vulnerable, eventually, to substantial operating costs. The result is a higher operating cost for the life of the infrastructure or equipment that far exceeds the front investment needed to avoid these costs. When staff members do manage to navigate these barriers to fund efficiency upgrades, savings generated often disappear into general revenue. Sometimes these managers then find the next year that their annual budgets have been reduced to reflect projected reduced operating costs. Clearly, this acts as a powerful disincentive for finding more efficiencies and sustainability-related initiatives.

In an effort to drive down staff costs, many colleges exist with such low staffing levels that there is little time available to seek out operational and system improvements and efficiencies. Capital project planning, for instance, can benefit from integrated design teams that carefully analyze the effect of design decisions on building performance, yet the failure to create sufficient time early in the design process for staff and consultants to evaluate options fully comes at enormous costs over the life of the building. Bringing in buildings on time and on budget is important, but that should not be used as an excuse to exclude thorough examination of more cost-effective and environmentally sound options that typically improve the comfort and productivity of building occupants. The product of all of these institutional finance and accounting processes can be a substantial hidden cost resulting from a loss of engagement, creativity, and innovation. By reforming their finance and accounting structures, organizations can establish new financial incentives and provide access to capital for more people, allowing for substantial financial returns to be generated, captured, and reinvested into a new green campus economy.

The Benefits of Campus Capacity Building

The speed at which a college or university advances toward sustainability objectives will largely be determined by the speed at which its employees and students can expand their understanding, personal commitment, and professional capacities to address the changes that will be necessary. The early champions who are quick to take action should have their successes widely promoted in order to motivate more hesitant colleagues. Eventually, as pilot projects and programs prove to be successful, efforts to institutionalize new practices can be accelerated by integrating new green workplace requirements in performance review cycles, including them as factors for bonuses and promotions. Performance review and advancement processes can include the completion of training programs that will develop new competencies in everything from green building design and operations to green custodial practices, waste reduction, and purchasing.

Routine assessments of emerging training and education needs should be conducted, and timely training should be provided for priority needs. A variety of immediate training activities should be instigated. Training on transformation leadership and change management for sustainability should be offered as well to the president and board of trustees. Enhanced meeting facilitation, negotiation, and team management skills should be provided to all managers, and training on life-cycle costing should be provided to the members of the financial office.

Establishing Change Management for Sustainability

The most important way for a college or university to launch itself on the journey toward campus sustainability is to establish a dedicated and effective sustainability change management function. One of the most important lessons learned since the 1990s about the greening campus movement is that without having an effective sustainability change management function, the institution will most likely struggle to make substantial ongoing progress. The work of enabling the entire organization to achieve continuous progress toward sustainability is just starting to be understood. At its heart, it is about making change easy, desirable, and rewarding so that large numbers of people across the organization set about fostering new innovations, practices, and behaviors in both the small and large arenas of their workplace. It also involves leveraging little victories to open the way for systemic transformation, all within a framework of good business practice and medium-term cost effectiveness.

With the integration of an effective change management team, the university or college can literally become good at change. This happens as a result of removing a variety of institutional and cognitive barriers and introducing incentives and support mechanisms that energize, inspire, and engage the workforce. Staff members thrive on stable experiences of change, and thus stable change is the essence of effective organizational change management for sustainability goals and objectives. The first order of business for the sustainability change agent is to set about engaging other people to become champions within the workplace. Eventually, he or she must grow the numbers of people involved across the organization until almost all are engaged as participants, facilitators, and leaders of stable processes of change. There are hundreds of new activities and roles for people across even a small campus infrastructure, ranging from swapping light bulbs to tracking greenhouse gas emissions and from reducing waste at events to reducing pesticide use on lawns.

The Challenges in Measuring and Reporting Effectively

Improving the triple bottom line of economic prosperity, environmental quality, and social well-being is a worthy goal but a very challenging metric. Easy to understand and widely supported goals and objectives, frequently articulated by presidents and provosts and made relevant to people executing daily tasks, are essential to the accomplishment of institutional change. Only by setting a course, monitoring trends, and achieving buy-in can large universities, for example, evolve purposefully. Yet developing performance standards for assessing the adoption of sustainability by higher education institutions remains difficult. Does an academic dean measure what students learn and whether they can apply their knowledge? Does the chief finance officer measure the way faculty time and research dollars are allocated? Or does the sustainability director measure how efficiently the institution uses energy, water, and materials or how it plans for and communicates its goals for the future and how well it enables its staff to attain those goals?

The first widely accepted metric to measure progress toward adopting sustainability across an institution of higher learning was introduced in 2009. The Sustainability Tracking, Assessment, and Rating System (STARS), developed by the Association for the Advancement of Sustainability in Higher Education (AASHE) and profiled in chapter 4 of this book is a tool to track progress in education and research, operations, and planning, administration, and engagement. The beauty of the tool is that it is so comprehensive. The challenge of the tool is the level of detail sought, the aspirational nature of some of the performance targets, and the subjectivity associated with some of the credits.

Making the STARS tool meaningful will require support from the top, automated reporting structures, and a central data repository visible to the entire campus community. Institutional research and sustainability offices are logical homes for this data. Monitoring progress over time, however, will require the involvement of departments across campus. Few universities, for example, currently flag the courses that incorporate sustainability principles or the funded research that advances sustainability. Developing a campus-wide baseline sustainability assessment and tracking progress over time will indicate where and when corrective actions are needed.

The Role of Human Resources

Since sustainability is a relatively new field, there are likely to be many current employees who are not familiar with the concept. If an institution is committed to advancing sustainability, then all employees need to know what that means for them and how they are expected to contribute to the effort. In-house training programs can be developed by human resources staff, by sustainability offices, or by outside consultants. As demand grows and budgets shrink, third-party trainers, webinars, and online tutorials are becoming more common. However, it is important to use institutionally specific examples of the concept being taught whenever possible.

Because adopting and applying a sustainability mindset is a learned practice, like recycling or yoga, it is helpful to institute frequent trainings and reminders. A growing number of campuses are employing peer-to-peer education to encourage the adoption of sustainable behaviors. Typically, the sustainability office and its campus partners develop the curriculum and then recruit departmental representatives who are willing and able to take the message to their colleagues. This ensures that a consistent message is delivered in a culturally relevant context.

Particular skill sets increasingly sought by universities and colleges interested in advancing sustainability include the following:

—Energy management

—Waste and recycling management

—Water, storm water, and wastewater management

—Building science and automation controls knowledge

—Procurement knowledge that includes familiarity with third-party sustainability standards

—Ability to inventory greenhouse gas emissions and develop emissions reduction plans

—Transportation demand management

—Sustainability coordination across the institution

—Public service and community engagement management

—Academic expertise in environmental economics, social justice, climate policy, and sustainability principles

Recruiting for Sustainability: A New Aspect of Search

MICHAEL A. BAER AND SEAN FARRELL

The dedication of an increasing number of colleges and universities to sustainability goals has been both a bottom-up and a top-down movement. Often it has been initiated by student and faculty concerns, but since 2005 or so, we have observed an increasing number of presidents, provosts, and trustees becoming engaged in these issues as well. Going forward, if we are to see widespread implementation of sustainable curricula and activities on campus, it will require leadership from chief executive, chief financial, and chief academic officers, who have the authority and resources to implement decisions involving budgets and personnel in the management of their institutions.

(*cont'd*)

Recruiting for Sustainability: A New Aspect of Search

This, then, raises the question of how higher education institutions are going to identify the right leaders to move and sustain future green initiatives. To date, even when we have listened to presidents and deans on various campuses talk about their commitment and dedication, we have not witnessed a pervasive emphasis in those communities regarding the hiring of emerging leaders based on their records of past support for or implementation of sustainability initiatives. To ensure that our campuses will take a role in developing a workforce that will contribute to research and service leading to the development of more sustainable ways of living and working, colleges and universities need to make a conscious decision to recruit individuals who are knowledgeable and supportive of sustainable thinking and mission development. We outline below actions for specific moments in the search process in which a strong focus can be placed on sustainability:

1. *The search committee membership delivers a message.*

One of the surest ways to embed sustainability values into a search is to assign a committee member, or multiple committee members, who have demonstrated a deep commitment to sustainability and a genuine knowledge of its role in university mission and operations. Much as we cannot imagine an institution striving for racial and ethnic diversity not including persons of color on its search committees, it is now hard to envision not including green advocates on search committees. Often, the college or university's sustainability director is well positioned to fulfill this role.

2. *The charge to the committee sets a direction.*

The committee will usually be given a specific charge that provides the formal guiding principle for the search. We recommend that the charge specifically reference the importance of identifying candidates who will support the institution's goal of promoting sustainability and that the search itself be conducted in an environmentally responsible way. If a search firm is to be retained, the committee can seek answers in their proposal or presentation to one or two very specific questions aimed at soliciting how deeply the firm is engaged in environmentally sustainable practices.

3. *The position profile communicates an institutional path.*

Another key early moment during which a focus on sustainability can be interjected is in the position profile. This document supplies the search committee a clear opportunity to communicate to potential candidates the value that the campus places on sustainability. It is a document that is typically read by many, not only by the successful candidate but also by others who may be conveyers of the goals of the institution in the broader higher education community.

4. *Committee interviews test the candidate's philosophy.*

At this point, we encourage our committees to ask each candidate what he or she has accomplished in previous positions rather than simply posing hypothetical questions about what he or she would do in the role connected to the interview. This is a moment to assess candidly the individual's record. How active and involved has he or she been in creating a sustainable community? We suggest the committee follow up by asking the candidate to discuss specific actions his or her campus may have taken in this area, what role the candidate played in them, and what the final assessment of their success or failure was. Even if the candidate is not selected, the individual will leave the process with a clear message about the recruiting institution's commitment to sustainability.

5. *The campus visit becomes a showcase.*

While many savvy candidates will have done extensive research before setting foot on a given campus, we still recommend to our clients that they provide a "bundle" of information for each candidate invited to campus. To the degree that sustainability is a major value for the campus, we would urge the campus to send material about the college's sustainability efforts as part of the package. Also, we would encourage campuses that are serious about a sustainability focus to be sure that some portion of each candidate visit incorporates this emphasis, particularly on the campus tour and in discussions with current students who are often viewed as the most authentic spokespersons for a sustainability-driven mission.

6. *Welcome the successful candidate and chart the path for success.*

Some administrators, presidents, and vice presidents for academic affairs have chosen to embrace the issue of sustainability from the moment they are offered the position. Whatever a new administrator's perspective on this, we encourage our clients to set up a Transition Working Group to help plan for and assist with the new administrator's arrival on campus. Again, we recommend that the institution's sustainability director serve on this group. Members will often commission or prepare white paper reports on the status of each operational area, including any short-term "hot issues" that are likely to land on the new person's desk soon after arrival. This is another excellent opportunity to reinforce sustainability thinking as a key institutional goal.

Michael Baer is a vice president at Isaacson, Miller, specializing in higher education. Sean Farrell is a managing associate at Isaacson, Miller.

Desired sustainability competencies related to a specific position can be identified prior to initiating the hiring process. If the hiring supervisor is not familiar with the sustainability aspects associated with a given position, then assistance could be obtained from the human resources department or the sustainability office. To ensure that all employees are working to advance sustainability, performance reviews can assess knowledge levels and routine behaviors. Just as sustainability is a process of continuous improvement, employees need to regularly update their knowledge and then modify their actions. Although many employees may jump at the opportunity to incorporate sustainability principles into their jobs, others will not act until it is clear that performance evaluations reflect this new institutional priority.

New Options for Performance Guidelines

Even small, resource-challenged institutions no longer have to go it alone when trying to incorporate sustainability programs and personnel. Third-party performance standards and certifications from reputable organizations can now provide industry-accepted guidance. One of the best-known certifications is available from the U.S. Green Building Council. Their Leadership in Energy and Environmental Design (LEED) Guidelines were developed and refined over multiple years by building owners, architects, engineers, product manufacturers, and professional associations. The emergence of LEED-accredited professionals and LEED-certified buildings has resulted in a shared vocabulary to discuss building performance targets. Instead of trying to explain each of the innovative features incorporated into a building design, college and university representatives can mention the number of LEED-certified buildings on their campus, and community members understand what that means. Most importantly, institutions and contractors tend to achieve a higher level of accountability when incorporating audited third-party standards. The LEED certification process is discussed in detail in chapter 13 of this volume.

Some third-party certifications assess product attributes; others assess processes. Green Seal, for example, rates cleaning products on the environmental impact associated with the product's life cycle. Cradle to Cradle evaluates a range of fabricated products from the molecular level to the degree product components are reused after their first intended life. The Carpet and Rug Institute assesses both carpets and the products and methods used to clean them with a focus on indoor air quality. The Forest Stewardship Council assesses how the trees used in wood-based products are grown, harvested, and processed. Energy Star certification, from the U.S. Environmental Protection Agency, indicates that a product or building is among the most energy efficient in its class.

Professional trade association guidelines, and building and plumbing codes, have long specified minimum accepted standards. The rate of incremental improvement required by these guidelines is increasing. The American Society of Heating, Refrigerating, and Air-conditioning Engineers (ASHRAE), for example, is now driving significant improvements in energy efficiency, indoor air quality, and the use of renewable energy technologies. Guidance is also provided on how to model projected building performance. Encouraging staff members to participate in these professional associations can influence the rate of continuous improvement and ensure that employees are on the leading edge of industry practice. It should be added that third-party certifications are increasingly viewed as points of departure, rather than as end points. Attaining certification indicates that a minimum level of performance has been achieved. For universities and colleges proactively seeking to mitigate their environmental impact, such certifications increasingly represent the least common denominator and a challenge to do better. Periodic evaluation of standards, tools, and guidelines is needed to ensure their effectiveness in aiding progress.

Reforming Our Systems

Lack of knowledge and outdated performance specifications underlie many current system dysfunctions on college and university campuses. Building design, construction, and management is rife with outdated policies, codes, and procedures—in part because this market has recently evolved so quickly after a long period of relative stagnation. For example, the technology exists to model new building performance and order construction materials directly from a three-dimensional digital model. Yet many higher education planners require paper design submittals, a resource-intensive process that results in less rigorous and accurate reviews. Investments in more efficient buildings are also hindered by the rigid divide that often exists between capital and operating budgets. Long-term utility savings and productivity benefits are frequently discounted, especially at the value engineering phase.

Purchasing is another area where higher education practices tend to lag significantly behind market developments. Tracking software, for example, can inform purchasers where products were grown, made, or extracted. Yet despite the potential to stimulate sustainable local economic development and reduce vehicle miles traveled, many purchasers do not consider the physical source of the product. Reviewing and enhancing building design guidelines and purchasing protocols is required on a regular basis. Developing schedules for reviewing and updating standards can remove the frantic urgency that too often accompanies contractual agreements and renewals. Monitoring trade journals, legislation, and regulatory actions can also help college planners to remain current and spot new trends.

Multidepartment teams are increasingly necessary to review standard operating procedures because no one individual can monitor all relevant trends. Climate action plans, for example, commit an institution to reducing greenhouse gas emissions. Yet laboratory technicians may not be trained to consider energy efficiency when purchasing research equipment. Information technology specialists may similarly not be familiar with the U.S. Environmental Protection Agency's EPEAT standards that assess the environmental impact of electronic equipment. Integrating emerging standards and environmental accounting into institutional practices will take time. Much of the information needed to make smart decisions is not readily available and must be proactively sought. The carbon intensity, embodied energy, or hazardous materials content of marketed products are examples of knowledge that is not readily available in the marketplace. Proactively obtaining and acting on this information is requisite for colleges and universities interested in taking a leadership role. History shows that manufacturers will not start to report such information until it is either required or requested by its largest customers. The responsibility to integrate these concerns either needs to be vested in a central department with influence or distributed to each relevant department on campus. Mechanisms to hold these respective units accountable for assuming leadership are then an integral component of the process.

The Need for Clear Communication

The role of presidents, provosts, and deans is to articulate a vision and then provide the tools and motivation to achieve that vision. Sustainable policies, practices, and curriculum affect virtually all members of a college or university community. The time to make known the expectations of an institution is before prospective students, staff, faculty, and administrators even arrive on campus. Campus tours, recruitment materials, and university websites can be used to convey the message of sustainability goals and objectives to prospective members of a campus community.

Research shows that the best time to introduce behavioral expectations to new students and employees is during their first days on campus. Orientation programs that stress an institutional culture of sustainability convey behavioral norms. Thus, newcomers to an institution who are interested in fitting in and succeeding in a new organizational environment are likely to adopt sustainable behaviors such as turning out lights, recycling discards, and taking public transportation. As with any aspirational goal, it is important for all members of an institution to recognize how sustainability relates to their role in the organization. Faculty can be routinely encouraged to focus on desired learning outcomes for students. Planners and architects can be asked to provide natural and built environments that reduce environmental impact, improve human health and comfort, and restore natural systems. Maintenance staff members can be requested to monitor market readiness and then install innovative products and technologies. They can also be evaluated in part on the degree to which they advance energy, water, and materials efficiency across campus.

Websites, annual reports, listservs, and newsletters are now common communication tools, but the problem with these passive information sources is that only those who are already interested are likely to peruse them. Frequent presentations by institutional leaders and guest speakers can carry greater import. In fact, one of the most effective means of gaining people's attention is to convene gatherings where individuals with similar interests socialize around food, drink, ideas, and then action. Food, which can also be used to convey a sustainability message, is a strategic plus if budget allows. With the growing national interest in local and sustainably produced food, refreshments themselves can tell a story of economic vitality and environmental quality. Signage that highlights local farms connotes freshness, nutrition, local economies, and close connection to the land.

Inspiring and empowering people to care and act upon issues larger than themselves is essential to fostering a sense of stewardship for a campus, its community, and the planet. When employees, especially young staff members, feel that problems are beyond their reach and that their individual actions are irrelevant, they can be defeated before they begin. Yet when people believe they are part of a community that can accomplish significant goals, they are much more accepting of setbacks encountered along the way. Taking on the role of an innovator and leader can be risky, but progress only occurs when fear of failure cedes to the compelling need for change and the hope of creating a better future. Students, in particular, benefit from learning the context for effective decision making as they pursue their own goals.

Updating the Curriculum

As Cornell University president David Skorton has said, "Sustainability is no longer an elective. It is a prerequisite."[2] Few universities or colleges, however, currently require their students to take a course or demonstrate proficiency in sustainability. Many colleges are now beginning to develop sustainability focused and related courses. Still, only a relatively small number of higher education institutions have degree programs in sustainability: Arizona State was the first to offer a doctorate in the field. Integrating and diffusing sustainability across the curriculum is a challenge. Faculty members have traditionally been rewarded for working deeply into a narrowly defined field, yet sustainability is inherently multidisciplinary. Environmental studies and social justice courses tend to be the first to incorporate sustainability, with geography, anthropology, planning, policy, marine sciences, business, and economics coming next.

Faculty who are not familiar with sustainability concepts need to be encouraged and trained to introduce these ideas into their classrooms. At some schools, small grants are available for faculty to develop new lesson plans and curricula. At Northern Arizona University and Emory University, faculty from diverse disciplines collaborate to identify ways of introducing sustainability into virtually any field of study in higher education. The connections between sustainability and curriculum priorities are examined in depth in chapter 11 of this volume.

Developing a Collaborative Research Model

Solving sustainability's largest challenges will require expertise from multiple disciplines. Climate change, the inequitable distribution of wealth and health care, and addressing breakdowns in our global financial system will not be finally remedied by a single researcher.

The institutions that will succeed in grappling with big issues and securing large grants to establish new research centers will be those that foster a culture of collaboration, and assessing the skill sets possessed by an institution's existing faculty can help to identify the big picture issues that an institution may be best suited to address. Universities with strong engineering and materials science programs, for example, have an edge when working to develop the next generation of batteries or solar panels. Institutions with strong public health and finance expertise may be best equipped to develop new models of affordable health delivery systems for underserved populations. Community colleges with strong sustainable agriculture programs can educate the next generation of small-scale farmers interested in selling directly to urban consumers who want to purchase local food.

Mapping the correlation between major sustainability objectives and the existing and aspirational strengths of an institution is especially important in this era of declining budgets. Few colleges can afford to be good at everything. Strategic planning to develop existing and future strengths can position schools of all sizes to be more successful in their niches. At the same time, many research universities with their emphasis on new knowledge can sometimes struggle with how to integrate sustainability into their classrooms, labs, and publications. As a field, sustainability is often approached as a series of problems to be solved, rather than as the purview of a specific discipline. Breaching the silos in higher education to address big picture issues and achieve practical solutions is an ongoing challenge.

Fortunately, some funding bodies, like the National Institutes of Health, the Gates Foundation, and the MacArthur Foundation, are realigning their priorities to address these global issues and focus on serving what some have termed "the bottom of the pyramid."[3] The billions of people in the world without clean water and basic health care, especially women and those most likely to be displaced by climate change, have not traditionally been the focus of academic research. That these funding bodies and others are starting to recognize and support these imperatives is indeed a positive trend.

Engagement Is the Goal

Colleges and universities that proactively adopt sustainable business practices and high-performance infrastructures can also serve as living and learning models for the physical and virtual members of their campus communities and beyond. The University of North Carolina system, for instance, explicitly focuses on the current and future needs of its state and works to align its teaching, research, and operations to help serve those needs.[4] As a public institution, the system recognizes its responsibility to assist the state in delivering quality K–12 education; supporting economic development; providing good health care, particularly to underserved populations; and developing policies and strategies to protect and restore environmental systems. For the students participating, especially those who choose to become involved in some way in institutional planning activities, classroom learning and campus infrastructure reinforce each other. The buildings and grounds, energy and water systems, and waste management practices can teach as much as textbooks. Undergraduates who both study and help to adopt sustainable approaches emerge with more confidence and clarity as they take their learning into the larger world.

Local governments and small businesses and can also learn from the often innovative approaches adopted by these higher education institutions. Hundreds of colleges have now conducted greenhouse gas emissions inventories. Many have constructed best management practices for treating storm water, and a growing number are also installing renewable energy infrastructure to heat water and power buildings. By combining these new approaches with a rigorous evaluation of the lessons learned, these schools can help to inform and advance the adoption of more effective tools and techniques by municipal leaders and small business owners. Business schools that focus on the strategic market advantage of socially and environmentally responsible products and practices can provide support services to aspiring entrepreneurs as well. At the Kenan-Flagler Business School in Chapel Hill, the Business Accelerator for Sustainable Enterprise provides education, networking, and individualized mentoring to people trying to bring sustainable products and ideas to the marketplace.

Finally, on a growing number of campuses, the emphasis on sustainable entrepreneurship has spread beyond its business school to provide support to people in any field who have a potentially marketable idea. Since the 1990s, most new jobs have been created, not by down-sized corporations, but by smaller companies who have identified unfilled niches. An increasing number of these businesses are now addressing issues of sustainability, appropriately, as they hire a new generation of more engaged graduates into leadership roles.

While there are a growing number of helpful resources available to guide the way toward a more sustainable future, becoming an effective leader for sustainability requires a personal journey, a personal commitment, and a unique dialogue with one's own organization. In this largely personal leadership journey, one of the greatest recurring challenges will likely be navigating the profound interdependence that exists between everything addressed in this chapter within an organization designed to function in silos, departments, and specialties, staffed by people whose identity is based largely around individual achievement.

Managing interdependence at this level of complexity requires a variety of new organizational conditions that enable a fast pace of learning and change for sustainability, including a new balance between collaboration and individual achievement, prevalent use of systems thinking and integrated design, bottom-up and center-out leadership working in concert with top-down authority, and a culture of learning and innovation. These organizational conditions encourage people to think beyond individual achievement, individual specialty, and individual department and to be in dialogue with the larger systems of which they are a part. Executive leadership for sustainability is therefore less about embracing a specific strategy and more about fostering a culture supportive of clearly defined organizational and behavioral change.

CHAPTER SIX

Sustainability: Shifting Definitions and Evolving Meanings

Davis Bookhart

> Whether our definition of sustainability is anthropocentric, biocentric, egocentric, ecocentric, econocentric, sociocentric, worldcentric, or perhaps simply personally eccentric, they are all valid.
>
> Defining Sustainability: A Hundred Perspectives

> I shall not today attempt further to define the kinds of material I understand to be embraced within that shorthand description; and perhaps I could never succeed in intelligibly doing so. But I know it when I see it.
>
> U.S. Supreme Court Justice Potter Stewart

Few words in the English language mean as many different things to as many different people as the word *sustainability.* On the surface, we have a good sense of what it means: goodness, fairness, environmental stewardship, and long-term value. It is the new "green" movement—environmentalism 2.0 with a twist—emphasizing a new progressive and inclusive social, economic, and environmental framework for society. In general terms, this all seems intuitive and favorable. In fact, the term sustainability has attracted the hopes and aspirations of many, representing an ideal just short of utopia. On many campuses the movement has been so successful so quickly that it has outpaced its own early definitions.

Largest Leadership Challenge in 2020

- We are seeing the infancy of sustainability efforts slowly make way for more mature programs characterized by stronger emphasis on planning and structure, dedicated and increasingly specialized staffing, and concrete goals tied to long-term strategic visions. A decade from today sustainability leaders may no longer be burdened by the task of continuously justifying a vision of sustainability; rather, they will be challenged with the Herculean task of implementing it.

Most Important Leadership Attribute

- Effective leaders of higher education will be those who demonstrate flexibility in adapting to changing conditions while remaining firm on the overarching vision of a strong, sustainable, twenty-first-century institution.

While higher education's view of sustainability often focuses on outcomes, a more subtle yet vitally important element of sustainability is change. At its core, sustainability is about transformation, about going from what we have today to what we hope will be tomorrow. Sustainability is not just the vision of what a sustainable world looks like but also the process of achieving it. Unfortunately, for many college and university presidents, provosts, and board members, positive impressions of a sustainable future do not translate well into practical steps for action. Sustainability is exhilarating in concept, but its multiple definitions are often too broad or shallow to be of much use to the leaders of busy campuses. For practitioner administrators, sustainability as a set of clear-cut campus actions has tended to defy simple or all-inclusive definitions.

The question, then, may be whether a definition for sustainability really matters at all. More important, perhaps, is developing a shared sense of familiarity and comfort with the general themes of sustainability, how they shape institutional mission and vision, and how they can influence higher education leadership strategies that comport with the larger, broader visions of a sustainable world beyond the campus. Instead of trying to identify and promote a single definition of sustainability, this chapter attempts to pull some of its larger thematic strings together to provide a more useful perspective for campus decision makers. These themes are interwoven in the majority of sustainability programs at community colleges, liberal arts colleges, and major research universities and form the foundation on which presidents and leadership teams can build the effective management strategies outlined in the chapters of this volume.

Sustainability as a Conceptual Framework for Institutional Leadership

Before exploring some of these various sustainability themes, it is instructive to take a step back to see how the sustainability movement and its core philosophy evolved. Traditionally, the environmental movement was largely ecologically based and primarily focused on flora, fauna, and all the pollutants that damaged them. Environmentalists promoted conservation and an appreciation of the wonders of nature but often had little to do with business management, product development, or other traditional means of incorporating environmentalism into the larger economic and academic frameworks of both higher education and society in general. Without those tools, the environmental movement was often in the position of demanding a halt to certain activities—logging, fishing, energy production, resource extraction—without identifying and promoting positive alternatives. When alternatives were offered, they often seemed disingenuous (e.g., "Find another career") or involved sacrificing both convenience and comfort. There were attempts to develop new environmentally friendly products, but too often they tended to be inferior

Three Emerging Trends

- *More focus on process.* Instead of small demonstration projects with large PR value, the trend is moving toward long-term strategies and goal setting.
- *More focus on campus community integration.* Greening will not work with just the greenies. The level of sustainability is directly correlated with how intimately employees embrace their role as drivers of the process. The trend is moving toward making every job a green job.
- *More focus on sustainability support services.* As more people assume responsibility for their piece of the broader sustainability goals, the trend is for sustainability offices to transition into providing support, training, and guidance.

Three Best Practices

- Annual progress reports—not fluff pieces, but using actual trends and metrics to measure progress
- Continuous and timely reevaluation of policies and standards to ensure that they reflect the priorities
- Continuous educational opportunities for staff to challenge their thinking and provide new perspectives on intractable problems

in quality and more expensive. The environmental movement continued to meet stiff opposition when it seemed too indifferent to the concerns of American corporations. Debates became polarized; many were forced to choose between the false dichotomy of being either for economic growth or for the environment.

The international development community was the first to recognize the need for a better way of addressing environmental issues. The environment versus economy debate might have been a good source of academic discussion on campuses in developed countries, but it was a matter of life and death in developing nations. Many regions were reaching a crisis point. While Western countries had established a model of allowing a certain amount of pollution as an acceptable cost for economic prosperity before finally starting environmental mitigation efforts, this model was not working in many developing countries where pollution levels were so high that they were actually inhibiting economic growth.[1]

In 1983 the United Nations launched the World Commission on Environment and Development (WCED) to address this challenge. The commission, chaired by former Norwegian Prime Minister Gro Harlem Brundtland, explored efforts that would "lead to the achievement of common and mutually supportive objectives which take account of the interrelationships between people, resources, environment and development."[2] The WCED, now known as the Brundtland Commission, was the first to articulate a definition of sustainability based on a balance between the economy and the environment. This definition is still operative today as a conceptual anchor grounding all contemporary efforts to further refine the vision of sustainability. The Brundtland Commission developed this succinct explanation:

"Sustainable development is development that meets the needs of the present without compromising the ability of future generations to meet their own needs."[3]

A Decision One Would Make Differently with Hindsight

- Sustainability is all-inclusive, but staffing is not. It is better to prioritize and target the use of limited staffing and resources than to spread them broadly and less effectively.

Single Most Important Piece of Advice for a New Professional

- Become a bridge builder; pay less attention to marketing, PR, and splashy visibility projects and focus instead on how to engage people through building strong relationships and identifying ways to connect with each individual.

This definition has stood the test of time because of its clarity of vision and simplicity. By bringing future generations into focus, the Brundtland definition cleared the path for thinking about how we live today as a precondition for how we live tomorrow. As a general rule, it has proved to be brilliant. However, as a practical matter for, in our case, college presidents and trustees, it can be troublesome. What exactly will future generations need? Should campus planners and architects assume, for example, that the energy crisis will gradually be resolved and that we will have no need for petroleum? Can we deplete more than our share of limited resources in order to build an infrastructure to help future generations do without them? Conversely, if we save limited resources for those future generations, will they then be similarly burdened to save those same resources for their progeny, and so on ad absurdum? Even with the difficulties in using its definition as a practical roadmap, the Brundtland definition began a shift away from traditional ways of thinking about economic development and the interconnections between progress and the environment. Until the 1980s, it had been common to think of the environment and economic development as a balancing act and to assume that leaders could have one or the other but not necessarily both. The Brundtland definition challenged that conception by suggesting that the balance is not between economics and the environment but between the present and the future.

Sustainability as a Target for Social, Environmental, and Economic Activities: The Triple Bottom Lines

On a more immediate level, college and university administrators have struggled for two decades or more to evaluate progress in meeting institutional sustainability goals. The Brundtland definition offers a long view, but what were some basic, clear-cut measurements that would confirm that the institution was heading in the right direction? The business community, already accustomed to benchmarking its success by its profitability, realized that similar metrics could also benchmark success from a sustainability perspective. Instead of the standard "bottom line" as the ultimate measure, perhaps there were additional bottom lines that measured sustainability success. Some began to employ a "triple bottom line" that evaluated the impact on a business's employees and the environment in addition to its profitability. This new emphasis on people, planet, and profit created a sense that businesses could be more successful when they focused on three pillars of performance instead of simply financial metrics. To put it another way, sustainability has become "the place where the pursuit of profit blends seamlessly with the pursuit of the common good."[4]

Higher education institutions and the nonprofit community have similarly adopted this model as a way to visualize how many parts can work effectively together toward a single mission. Instead of pure profitability, the emphasis focuses more broadly on prosperity and on more widespread benefits. However, the most important element remains sustainability as the context in which all three categories overlap (fig 6.1).

The thinking on many campuses now is that a healthy natural environment is not sustainable without positive economic growth, which, in turn, is not possible without a vibrant and equitable community of citizens and workers. The European Commission went further in 2009, saying that actions must regard "social, economic and environmental needs as inseparable and interdependent components of human progress" in order to contribute to an overarching vision of sustainability.[5]

A Sustainability Myth

- The myth is that campus sustainability can be accomplished by sustainability professionals alone. Sustainability cannot advance with only a handful of dedicated staff members. Sustainability must involve all members of the campus community.

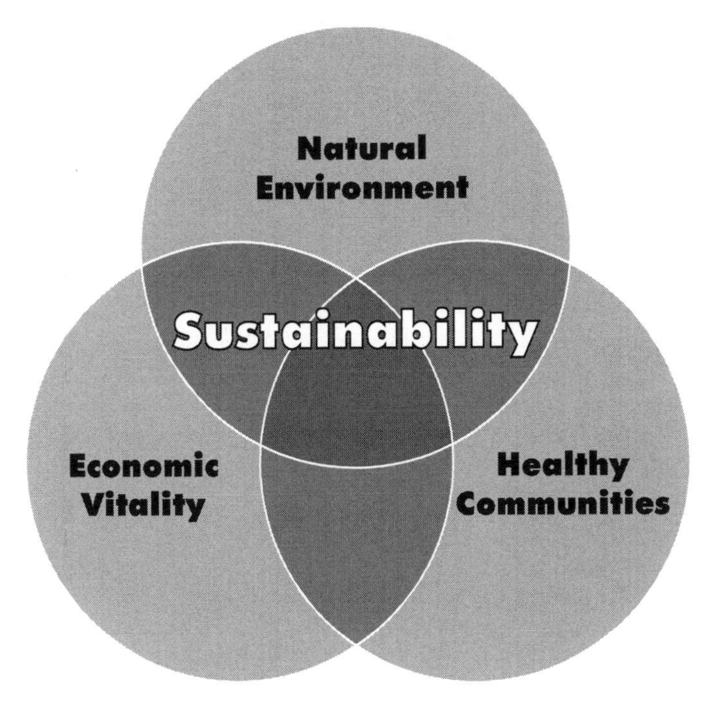

Figure 6.1 The space where these three circles overlap may represent the optimal area of opportunity for sustainability efforts.

The triple bottom line approach is very adaptable to the complexities of campus operating systems, especially in its capacity to incorporate a sense of process into planning and action. Now more pointed questions can be asked when building projects are developed or programming activities completed. Did this project produce positive environmental and social benefits in addition to being economically viable? What were the trade-offs? Emory University, for example, has developed a vision for action whereby "progress will be measured using the environmental, economic, and social 'triple bottom line' of sustainability."[6] When developing a long-term sustainability master plan for the City of Baltimore, the Baltimore Commission on Sustainability decided that all future actions and decisions involved with implementing the plan must be measured against the question of how they impact each category in the triple bottom line.[7]

Unfortunately, the approach is also static. It does not necessarily provide a perspective on the future; in fact, it may even undercut the long-term vision. Some decisions today, such as those in the climate change debate, simply may not have an obvious economic benefit and yet may retain significant ramifications for the future. Should restoration of the Everglades, for example, be abandoned because there is no obvious "people" benefit? Or, more relevant to universities and colleges, how do the costs and benefits of training, communications, and scholarship get evaluated within a triple-bottom-line framework?

Recognizing that presidents are often challenged to make difficult decisions based on the future benefits and not necessarily the immediate triple-bottom-line metrics, the Baltimore Commission on Sustainability attempted to reconcile this issue by combining the Brundtland vision with the triple-bottom-line approach. The definition became "meeting the environmental, social, and economic needs of Baltimore without compromising the ability of future generations to meet these needs."[8] An increasing number of community colleges, liberal arts colleges, and major research universities are pragmatically exploring the benefits of similar definitions that combine a focus on the future with these three benchmark categories of evaluation.

Sustainability as a Measure of Ecological Impacts: Moving from Definitions to Action

Going back to the roots of the environmental movement, an alternative view of sustainability is one that takes an ecological perspective—both physically and metaphorically—and translates that view into goals for the future. Defining sustainability in ecological terms may be one of the most challenging of all approaches for presidents, provosts, and trustees, but it also carries a high potential for supporting various programs and initiatives that are helpful in presenting the concept of a sustainable future to the broader community. As a metaphor, ecological sustainability paints vivid images of human impacts on the natural world. We tend to use the term "footprint" to measure the amount of damage we are causing on the earth, as if our institutions are literally stepping on, and blotting out, the ground under which we tread. We have carbon footprints, building and construction footprints, campus footprints, to name just a few. Footprints leave marks, and the image of trampled ground under footprints is an effective visual in terms of understanding the state of the natural world and the need for a sustainable approach to minimizing the impacts of our collective feet. As Ken Wilber posits, "Sustainable development is the lightest footprint possible on the biosphere and, most importantly, a consciousness that can sustain it."[9]

There are numerous ecologically based definitions, but most tend to come back to a central theme: "stabilizing the currently disruptive relationship between earth's two most complex systems: human culture and the living world."[10] The natural world, through its abundance of riches, has a remarkable ability to support human needs. However, there is a limit—a "carrying capacity" to the earth—over which human needs become unsustainable over time. It is clear that we are over that limit, so ecological approaches to defining sustainability tend to focus on rebalancing the natural systems within the framework of human activity. The Australian government's Department of the Environment, Water, Heritage and the Arts developed this helpful definition: "Ecological sustainable development is using, conserving and enhancing the community's resources so that ecological processes, on which life depends, are maintained, and the total quality of life, now and in the future, can be increased."[11] This definition shares a common theme with the Brundtland definition, namely, the focus on the future, but it emphasizes the intricate relationship between the natural world and human communities. The definition also underscores the vibrancy of sustainability and, for campus leadership teams, returns us to the point where this chapter began: Sustainability is change, it is a process rather than an end, and managing this process effectively is a necessary goal for the entire campus community.

It is in emulating the natural world that sustainability gains so much power as a concept. There is power in a president or chief academic officer asking her or his institution: How do ecological systems exist in ways that make them sustainable? What lessons can we learn from these systems in terms of how we deal with energy, waste, consumption of resources, interaction between and among species, and sufficiency? At Johns Hopkins University, the community begins with this question: What does a truly sustainable campus look like? While there are still few clear answers, the process of introspection framed by an ecologically driven definition of sustainability has proved to be an excellent launching point for consideration of both strategic and annual operating plans across the university.

Given the diversity of sustainability definitions both on and beyond the campus, it would seem that the term would remain "as illusive as any one of the slipperiest pieces of soap you are ever likely to find in the shower."[12] Regardless, the concept seems to be widely accessible and generally understood. All definitions seem to embrace one or more aspects of treating the natural world with deference, respecting the value of humans, and behaving—as groups or as individuals—in a responsible and accountable fashion. The definitions also have one more thing in common: an attempt to foster a vision for how things can be different next year and a generation from now. In the end, sustainability is a powerful concept precisely because it means so many different things to different people. At its core, sustainability is highly personal, and therefore it is accommodating as a common platform to help leaders shape their visions for the future. In this sense, it is the vision not the definition that is transforming higher education communities.

CHAPTER SEVEN

Sustainable Citizenship: The Challenge for Students and Their Institutions

Terry Link

> We stand at a critical time in Earth's history, a time when humanity must choose its future. As the world becomes increasingly interdependent and fragile, the future at once holds great peril and great promise. To move forward we must recognize that in the midst of a magnificent diversity of cultures and life forms we are one human family and one Earth community with a common destiny. We must join together to bring forth a sustainable global society founded on respect for nature, universal human rights, economic justice and a culture of peace.
>
> The Earth Charter Initiative

Sustainability has sometimes been defined as balancing a three-legged stool over the long haul. The legs represent the environment, society, and the economy. Higher education curriculum has traditionally favored a reductive approach to knowledge by having students focus attention primarily within one major area of specialization. Since the complex challenges facing the human family cut across a wide swath of knowledge areas, sustainable solutions will require collective action. These solutions, and the students learning about them, need what author Tom Atlee has called "co-intelligence."[1]

Co-intelligence is the synthesis we achieve when we provide the conditions for a group of people to bring forth their best thoughts in a collaborative problem-solving mode. Yet within colleges and universities, we create systems of "expertise" that certify young professionals largely by recognizing narrow research areas. This long push toward reductionism has costs as well as benefits. Let us consider a current, complex issue: climate change. An atmospheric chemist may argue that the concentration level of CO_2 in the atmosphere is creating global warming. Depending upon how she or he reads the data, an argument to reduce human-released CO_2 is the most prudent approach. On the ground, an agronomy expert might argue for an agricultural approach, concerned that changes in the chemistry of soils may have side effects that will create food supply issues. Meanwhile, an economist who does not calculate climate change and resource depletion into her or his equations may argue that preferred solutions are simply economic ones and all that is needed are the right incentives. With each expert arguing one worldview in a battle of wits, we fail to create the environment for co-intelligence to emerge.

While a doctoral degree has been blessed as the epitome of wisdom by our higher education model, it may also convey only a narrow and rigid understanding of this kind of knotted challenge. In fact, it is obvious to some that much of higher education culture continues to support a battle-of-wits approach over an inspired, deliberative co-intelligence. The Wuppertal Institute, with the European Friends of the Earth, has developed a model of sustainability that they call the "Prism of Sustainability."[2] This four-pointed tetrahedron adds a missing and, I believe, crucial dimension: democracy. This often-overlooked ingredient in many sustainability discussions on campus was also quite visible in the Earth Charter's idea of sustainability:

—Concern and Care for the Community of Life

—Ecological Integrity

—Social and Economic Justice

—Democracy, Nonviolence, and Peace

The inclusion of democracy as a crucial element of sustainability addresses two key questions at the heart of this chapter: who decides, and how are decisions made? This book addresses the role of higher education leaders, particularly presidents, provosts, and board members, in developing best practices for sustainability on their campuses. In turn, the emphasis in this chapter is on the specific roles and responsibilities their colleges and universities then hold in preparing their students to become engaged citizens as well as in educating the rest of each campus community to awaken to their roles as citizens of a more sustainable world. John Hannah, former president of Michigan State University and chair of the U.S. Commission on Civil Rights, is memorialized in a statue not far from my former office window. At the base of the statue are inscribed these words: "If there is one thing educators agree on, it is that the principal role of education is to create good citizens." In the twenty-first century, what does this require?

The Challenge for Institutional Leaders

My generation grew up with the boundaries of citizenship defined as those of nation-states. We have learned more recently through the lens of sustainability that the many boundaries that separate us into political jurisdictions are arbitrary. From the local to the global, we traverse these boundaries regularly via travel, commerce, and culture. With the realization that we are so profoundly interconnected, what does today's notion of sustainable citizenship require of us—and of current students, even more critically? As we grapple with global climate change and dwindling resources, how shall we use them? What is a fair distribution among those living now? Arriving at answers we can all live with requires some framework or system that is fair for all and that affords everyone equal opportunity. For campus leaders such as presidents and provosts, then, what are the rules of engagement? In short, who decides?

In academic governance models, it is not uncommon to have faculty, students, and senior administrators all involved in some level of decision making, yet it is still quite unusual to provide support staff members a voice. During the more proactive affirmative action era, it was common in higher education to place a designated affirmative action advocate on its search committees to help ensure that various concerns were not overlooked in the hiring process. Yet how many higher education institutions today place an advocate representing future generations on committees where the decision has long-term impacts? How do our internal governance structures teach our students the democratic ideals we preach in Political Science 101? Concepts like proportional representation, instant voter runoff, citizen commissions, public budgeting, and various other experiments in effective governing continue in places where citizens want to create more fair and powerful systems of shared decision making. How well versed in these approaches are the undergraduates emerging from our universities and colleges? We offer hundreds of thousands high school civics classes that examine our current system of government, but rarely do our campus curricula explore the necessary structures of a truly democratic governing scheme or consider the wide array of emerging innovative practices and approaches to effective governing. More broadly, what experiences in the daily operations of a college nurture the engagement of students and their professors for the betterment of that community? The various chapters of this book focus on some of the best practices that key operational areas can model.

Successful Strategies for Student Engagement

Perhaps the most crucial role for undergraduate programs to play in this effort is in teaching and then ensuring student empowerment. The pedagogy of how this is accomplished and made real is centrally important to the undergraduate experience both in and out of the classroom. Simply teaching sustainable citizenship as a detached intellectual concept will end in failure. Rather, choosing approaches that engage learners, with sleeves rolled up, is key to how they ultimately perform their role as citizens in whatever communities they will call home. Just as teaching a student how to make a piece of furniture requires hands-on experience, so does the teaching of sustainable citizenship. What if every semester, in courses across the curriculum, students were encouraged, inspired, and required to connect community engagement efforts with the principles of their majors and disciplines? I have often asked students in courses and workshops at Michigan State University to develop a personal map in which they list or draw all the possibilities where they believe they have influence in their daily lives. Almost all participants list family first; other frequent listings include jobs, athletic teams, and student organizations. Some add churches, and a few mention government. However, higher education institutions continue to overlook the importance of helping students develop their abilities to make change beyond themselves. Pedagogies that still depend on a stand-and-deliver approach can fall short by continuing a stereotype that expects students to simply bow to an expert authority. Is it any wonder, then, that many students shy away or even rebel when a professor challenges them to be co-learners and to take some responsibility for shaping their own learning environment? This learned helplessness cannot inspire or produce lifelong sustainable citizens.

Engaging students directly in the issues confronting them and their communities while instilling the belief that they can make a difference is one of the core components of teaching sustainability thinking. Whether these undergraduates learned any guidelines in high schools or from their families—and many did not—colleges and universities still need to teach the tools of both global citizenship and local engagement. Regardless of the degree program one chooses, easy places to begin learning about sustainable citizenship, even as an undergraduate intern, are state arts funding initiatives, state and municipal tax plans, incentives

The Student Voice: Action Steps for Sustainability Success on a University Campus—The University of California

MATTHEW ST. CLAIR

In fall 2002, students from the ten University of California (UC) campuses demanded that the UC Board of Regents adopt clean energy and green building goals. By the following spring, they had convinced the Regents to do so.[a] Students have increased their advocacy over the years, successfully working with UC leadership to expand UC's Policy on Sustainable Practices to all other areas of sustainable business practices.[b]

The top tier of every campus sustainability ranking to date has included at least one University of California (UC) campus or the university system as a whole. These rankings are a recruiting tool to attract top students, and the example above demonstrates how students themselves contribute to the sustainability performance that earns UC campuses these accolades. When the Princeton Review started including a sustainability question in its annual survey of 10,000 college-bound high school seniors, more than 60 percent of respondents said their choice of college would be influenced by how sustainable the campus is.[c]

The following are three suggestions to ensure that a university campus benefits from student energy focused on sustainability issues:

1. *Hire a sustainability officer who works well with students.* An effective sustainability officer can transform a potentially adversarial relationship into a productive one that catalyzes student energy. This requires willingness to demonstrate a public commitment to sustainability and is best accomplished through establishing a sustainability office and hiring a sustainability officer who can mentor students and create co-curricular educational opportunities.

2. *Create a sustainability internship program.* The chancellors' offices at UC Santa Cruz and UC Berkeley fund sustainability internships that attract top students into prestigious, paid positions.[d] These are typically sponsored by campus departments that have sustainability projects that students are uniquely positioned to carry out.

3. *Create a sustainability grant program.* The UC Berkeley chancellor allocates funding each year for five to eight Chancellor's Green Fund Grants.[e] These grants legitimize and recognize innovative ideas that require a limited amount of funding to implement. UC Berkeley Chancellor Robert Birgeneau calls the Green Fund "the best investment that California Hall makes. The return on the investment is something like a hundred to one."[f]

(*cont'd*)

The Student Voice: Action Steps for Sustainability Success on a University Campus—The University of California

Ultimately, sustainability leadership is not just something that students want from administrators; it is something that students themselves are providing. Every college and university will benefit to the extent that it can effectively mentor and stimulate that student leadership.

Matthew St. Clair was hired as the first Sustainability Manager at the University of California's Office of the President in 2004 after, as a UC Berkeley graduate student, he helped lead the student campaign that convinced the UC Regents to adopt a sustainability policy in 2003.

[a] www.universityofcalifornia.edu/sustainability/documents/regpolicy.pdf.
[b] www.universityofcalifornia.edu/sustainability/documents/policy_sustain_prac.pdf.
[c] www.princetonreview.com/uploadedFiles/Test_Preparation/Hopes_and_Worries/HopeAndWorries_Full%20Report.pdf.
[d] At UC Santa Cruz, the Office of Sustainability has helped campus departments create sustainability internships through the already-established Chancellor's Undergraduate Internship Program, taking advantage of a preexisting program and structure: http://intern.ucsc.edu/cuip/student.html. For Berkeley, see http://sustainability.berkeley.edu/cacs/pages/internships/overview.shtml.
[e] http://sustainability.berkeley.edu/cacs/pages/greenfund/overview.shtml.
[f] www.sustainability.berkeley.edu/cacs/summit.

for small businesses, health regulations for local communities, and, of course, environmental policy development on a national scale. Beyond the curriculum, a lifelong opportunity for students and alumni to direct their extended communities toward sustainable goals is the consumer marketplace. Too often, college and university leaders overlook the power their communities can collectively exert through spending choices, product preferences, and direct calls for changes in corporate policies. These consumer expectations drive world markets, and even media-saturated first-year communications majors sometimes fail to realize that each dollar spent is a vote affecting the marketplace—whether for change or the status quo.

Additionally, a large number of faculty and administrative leaders in American higher education are involved in progressive models of sustainable citizenship beyond their institutions, but these activities are not always translated into road maps and simple best practices for their students. As we work to integrate those initiatives with the growing number

The Student Voice: Action Steps for Sustainability Success on a College Campus—Berea College

FINNLEY HAYES

At Berea College, many entry-level courses focus their content on the environment from perspectives such as environmental degradation, bioremediation, and peace and social justice. Other General Studies courses use this same approach by using environmental issues as the backdrop to everything they teach. In these ways, Berea's courses attempt to lay a foundation to help students think about their present and future roles in the world beyond our campus.

The following suggestions have worked well at Berea to engage students in sustainability thinking and action:

1. *Facilitate direct student involvement immediately.* As one example, Berea hosts competitions among its residence halls to determine which can use the least amount of resources. An effective way to accomplish this is to install monitors in student lounges and hallways. These devices will collect the figures for a day's electricity usage and compare it to the average for the week before: if it is higher, the monitor will register red; if it is lower, it will register green. Building on this strategy, discounts or rebates might be offered to future students who are willing to adjust their consumption habits accordingly.

2. *Host regular campus-wide events supported by adequate budget resources.* Multiple student organizations at Berea emphasize the stewardship of the earth. Colleges should integrate these organizations into their budgets and encourage group and club leaders to host as many trips off campus as funds will allow to tour destinations such as "green cities" and mountaintop removal sites in order to make students aware of what is happening beyond their classrooms.

3. *Implement an interdisciplinary major.* In terms of curriculum, colleges should consider implementing, at the least, an interdisciplinary major titled Environmental Studies. First-level courses in this major can identify challenges and breakdowns in our current lifestyle choices, and upper-level courses can develop and examine solutions to these concerns. All majors should offer senior seminars and capstones with as much flexibility as possible for students to design projects that support career commitments and special interests.

Finnley Hayes is a member of the class of 2012 at Berea College, majoring in Ecological Design.

of degree programs specifically focusing on sustainability-related career choices, institutional "outcomes," not those simply earned by students, will reach higher levels. Finally, this is where presidents, provosts, and senior administrators need to direct new attention and resources. Regardless of whether a student graduates in botany or business administration, he or she will become a citizen of some place, and that place will be inextricably connected to other places. Making those places more enriching and meaningful communities is one of the most important goals for present and future students as sustainable citizens, and they need our help to achieve it.

II

Sustainability and the Leadership Team: New Assignments

CHAPTER EIGHT

Sustainability and the Presidency: Five Starting Points

Jo Ann Gora and Robert J. Koester

The challenges of sustainability are at the collective doorsteps of American campuses. As discussed in the preceding chapters, there is a context and a history to that arrival and—as will be presented in this and the chapters that follow—there are many potential scenarios for an as yet untold future. This chapter, however, is intended for presidents and board chairs who have chosen to embrace the leadership opportunity and want to kick-start an action agenda on their campuses.

Each senior administrator operates from a different perspective within institutions of differing history, operational complexity, and community connections, so our respective points of departure will necessarily be unique. Nonetheless, for each president, trustee, or even provost, the healthiest leadership approach will be one that is holistic. The presidency deals with such a broad array of constituent groups and such wide-ranging scales of interaction that finding an effective framework by which to steer through these challenges is critical. Although at first this may seem a daunting operational task, the complexity and dimensions of this opportunity can, over time, align comfortably with the interests of those with whom we work and whom we serve.

If we are new to our presidential post, we are faced with understanding more fully the complexities of the institution we have joined so as to develop strategies appropriate to helping that institutional culture make its way through the transformation to sustainability. If we have held authority for five years or more and these issues are finally reaching a level of broad-based intense concern, the challenge is no different. We must look for those models that best fit our current operational practices to help nurture the needed change. What follows is a suggested, and tested, way to begin.

There are five specific starting points that can serve as a framework for this leadership work, and although we will frame these starting points in a sequence of presentation, we recommend initiating all five simultaneously. Specifically, we advocate the following: (1) staying current with international, national, regional, state and local developments; (2) taking a "whole-systems" approach to the action opportunities; (3) creating a campus-wide culture of collaboration and innovation; (4) using multi-modal leadership techniques to stimulate the needed transformations; and (5) socializing the board of trustees, who all too often have not had the time to become informed about the issues and who therefore look to presidential leadership for guidance on how to support an institution-wide leveraging of the opportunities at hand.

Staying Current

Monitoring the national higher education landscape is fundamental to becoming more fully informed about the opportunities for presidential leadership. Many of the individual stories and collective trends of institutional achievement are reported routinely through the *Chronicle of Higher Education* and, perhaps most importantly, annually in the AASHE Digest. However, some of the more immediate and fruitful sources of information can be found in the numerous email list servers that have arisen within and around the many formal operational organizations that comprise the professional interest communities of higher education (see table 8.1).

In fact, a recent development includes organizations that are themselves acronyms of acronyms: the Disciplinary Associations Networks for Sustainability (DANS) and the Higher Education Associations Sustainability Consortium (HEASC). DANS is focused mostly on the academic side of the higher education enterprise, and HEASC on the business affairs side. As an example, the DANS mission statement reads:

> We, the members of the Disciplinary Associations Networks for Sustainability (DANS), declare our commitment to education for a sustainable future. In response to the planetary challenges of the 21st century, we believe that the engagement of the academic disciplines is critical to advancing the broad goals of sustainable development. We seek to help higher education exert strong leadership in making education, research, and practice for a sustainable society a reality.[1]

Similarly, in its mission statement, HEASC writes:

> We, the members of the Higher Education Associations Sustainability Consortium (HEASC), an informal network of higher education associations,

Table 8.1. National and International Sustainability Organizations in Higher Education

Association for the Advancement of Sustainability in Higher Education (AASHE) www.aashe.org/
Association of University Leaders for a Sustainable Future (ULSF) www.ulsf.org
Disciplinary Associations Network for Sustainability (DANS) www2.aashe.org/dans/
Higher Education Associations Sustainability Consortium (HEASC) www2.aashe.org/heasc/
National Wildlife Federation (NWF)—Campus Ecology Program www.nwf.org/campusecology
Second Nature www.secondnature.org
U.S. Partnership for Education for Sustainable Development— Higher Education Sector www.uspartnership.org/main/show_passage/54

> affirm our commitment to advancing sustainability within our constituencies and within the system of higher education itself. Our current member associations that make up HEASC see the need for developing in-depth capability to address sustainability issues through our associations and have decided to work together in this effort. We, the members of HEASC hope to involve *all* higher education associations to get the broadest perspectives and produce the greatest effectiveness and synergy in our efforts.[2]

In addition to these national initiatives, much regional collaboration has developed throughout the country (for a list of these regional sustainability organizations, see the appendix at the end of this chapter). More specifically, for those of us in Indiana, numerous coalitions and groups are actively working to bring sustainability to the educational experience of students, to the operational policies of institutions, and to the public commitments of college and university leadership (table 8.2).

One of the more profound early developments on the international stage of sustainability in higher education was the 1990 creation of the

Table 8.2. Notable Indiana Sustainability Organizations

Earth Charter Indiana www.earthcharterindiana.org/
A Greener Indiana: Indiana's Sustainable Agriculture www.agreenerindiana.com/group/agriculture
Hoosier Environmental Council www.hecweb.org/
Indiana Campus Sustainability Alliance—Uniting Indiana's Campus Community to Promote Green Policy and Practice www.indianagreencampus.org/
Indiana Office of Energy Development www.in.gov/oed/
Indiana Renewable Energy Association—Promoting Environmental Sustainability in Indiana www.indianarenew.org
Sustainable Indiana www.sustainableindiana.org/
Sustain Indy: Growing a Livable City www.sustainindy.org/
Windiana: The Indiana Wind Working Group Conference www.cec.purdue.edu/regforms//Wind09RegistrationForm.pdf

ten-point Talloires Declaration as a framework for institutional transformation. Written by Anthony Cortese while he was serving as dean of environmental studies at Tufts University, the document was signed initially by thirty-five chancellors and presidents at the Tufts University campus in Talloires, France. To date, some 350 institutions have signed the document and have implemented some, if not all, of its tenets. Our own institution, Ball State University in Muncie, Indiana, signed it in 1999, and we have used it to frame our approach to organizing a campus administrative structure to engage the sustainability challenge. More recently, the United Nations declared 2005–2014 as the UN Decade of Education for Sustainable Development. As stated by the lead agency for the decade, the UN Educational, Scientific and Cultural Organization (UNESCO):

> The international community now strongly believes that we need to foster—through education—the values, behavior and lifestyles required for a sustainable future. Education for sustainable development has come to be seen as a process of learning how to make decisions that consider the long-term future of the economy, ecology and equity of all communities. Building the capacity for such futures-oriented thinking is a key task of education.[3]

Another recent example is the singular call for presidential leadership in the form of the American College & University Presidents' Climate Commitment (ACUPCC), which has well over six hundred signatories with the goal of acquiring the participation of all presidents and chief executive officers of colleges and universities nationwide. This program is administered and supported by the Association for the Advancement of Sustainability in Higher Education (AASHE), the fastest growing and most fully vetted operational and information resource for the many constituents that comprise every university community.

Another nationwide AASHE program, which was implemented in 2009, is the Sustainability Tracking Assessment and Rating System (STARS). Sixty-six colleges and universities participated in its pilot year to test this rating tool and provide initial feedback. The final version was introduced at the national Greening of the Campus Conference VIII in September 2009, co-hosted by Ball State. This rating tool organizes performance metrics under the categories of Governance and Administration, Community Outreach, and Research and Education. The intent is for universities to self-score their achievements in these areas and issue annual reports alongside those of other institutions.

Using a Whole-Systems Approach

Strategic planning is the watchword of any effective presidency, and the challenge is to weave sustainability content into that strategic planning while balancing both the centralized and the decentralized cultivation of engagement by faculty, staff, and students. At Ball State, we elected to *require* in our strategic plan that every administrative unit originate a sustainability plan using a template prepared by our Council on the Environment (COTE), discussed in more detail later in this chapter. The template was organized according to the framework of the STARS rating system, and units were asked at the turn of the year to file with the provost their five-year, unit-level sustainability plans. Responses were remarkably diverse. In one college, the office of the dean prepared a generic response that was then adopted verbatim by his respective department chairs. In the case of our Department of Technology, and more generally our College of Applied Sciences and Technology, the faculty and administrators engaged this opportunity with full force and developed highly diverse and rather sophisticated proposals. In the case of our Honors College, the students originated the plan. All of these are now available online at www.bsu.edu/cote/sustainabilityplans.

Part of any sustainability plan—whether centralized under presidential leadership or localized to an administrative unit—must be a curricular focus. In these days of interest in asynchronous distance education, many new educational options exist, from certificate programs for post-degree professionals to topical academic minors under the rubric of sustainability. These require, however, that champions for the options be invited to provide the leadership. Annual reporting on strategic plan outcome measures presents a tool for assuring such involvement. By identifying success stories and sharing them at the close of each academic year, we can foster replication by others during years that follow.

An additional incentive for cultivating faculty involvement within academic and staff units throughout the campus is recognition through a formal awards program. Our Council on the Environment (COTE) has instituted just such recognition. We identify "Green Initiatives"—actions taken by individuals or administrative units to advance principles of sustainability in their day-to-day practices and work. These initiatives can have modest or substantial impact; what is being commended is the initiative itself. We are trying to stimulate an acknowledgement by faculty, staff, and students that it is within their power to take action without waiting on formal organizational and oftentimes bureaucratic mechanisms being brought to bear. This awards program extends to the East Central Indiana community, and we have recognized a number of initiatives by individuals and groups from our Indiana citizenry.

In addition to the "Green Initiatives" recognition, COTE instituted a best-practices Exemplar Award and the Lifetime Achievement Award for contributions by individuals, groups, or administrative entities over time. Thus, we try to address the immediately doable short-term action as well as the longer-term building of a legacy of impact. These acknowledgments can include activities with specific academic focus. Cultivating co-curricular opportunities for students adds dimension to their collegiate experience and provides a foundation for advancing sustainability initiatives. Many universities have adopted Eco-Reps programs and hosted residence hall competitions for energy and water conservation and for materials recycling (as part of the national Recyclemania Program) or have encouraged students to work with local groups such as Habitat for Humanity, the Sierra Club, River Keepers, or similar organizations to implement sustainability practices as a hands-on experience.

On a broader scale, for thirteen years Ball State has hosted a biennial Greening of the Campus Conference from which we have garnered the experiences of universities throughout the United States and around the world. During 2009, the university decided to stage it at the Indiana Convention Center in downtown Indianapolis. Moving the event off campus allowed a larger gathering, simplified access for conference participants, and accommodated a Green Campus Exposition enabling vendors and nonprofits to display their services and good work.

We were joined by the Association for the Advancement of Sustainability in Higher Education (AASHE) as a full partner in the event this year. For the last several cycles of our conference offering, our biennial event has been occurring in alternating years with that of AASHE. Given the strength of their conference gathering last year and the growing numbers of attendees, it made sense to establish this partnership; in fact, doing so was in the spirit of the theme chosen for the conference: *Embracing Change*.

Finally, a whole-systems approach requires operating transparently with a consistency of communication. Posting on our web pages the unit-level sustainability plans, sharing publicly through the ACUPCC website our Greenhouse Gas Inventory (and ultimately our Climate Action Plan), and revealing our plans for decommissioning our coal-fired boilers and shifting to a geothermal district heating and cooling system reflect our commitment to such transparency by sharing publicly the means by which we are implementing the principles and practices of sustainability.

Moreover, our university's theme of immersive learning provides a mooring for a consistency of communication. All that we do is focused on the student experience; the means by which we engage these principles and practices of sustainability are framed from that point of view. In fact, we have tagged our whole-systems approach to education for sustainability as itself a form of *institutional* immersive learning. As an intentional community, we are engaged en masse in a continual *tracking* of history, *evaluation* of progress, *modification* of approach, and a *refocusing* of effort. In our day-to-day immersion in the complexities and challenges of sustainability, we are consistently *celebrating* achievements, *facilitating* transformations, and *anticipating* next steps in our continuing campus-wide community renewal.

Creating a Campus-Wide Culture of Collaboration and Innovation

Colleges and universities are often characterized as a collection of silos in which differing units have little or no interaction, much less collaboration. This is often true of relations between academic affairs and business affairs. Provosts and the senior faculty principally responsible for the academic mission of the institution often take for granted the financial management, operations, and facilities support by those responsible for business affairs. Conversely, those involved in the day-to-day facility management challenges can feel disconnected from the specifics of academic content delivery. Yet for an institution to engage substantially the tenets of sustainability, such implicit barriers must be erased. The opportunity for both teams to work together offers potential for spirited interchange, leveraging resources and creating a unifying sense of community for administrators, faculty, staff, and students alike.

At Ball State University, the president has viewed cultivation of a sense of entrepreneurial spirit throughout the institution as a key leadership responsibility. Students, staff, faculty, and administrators are encouraged to act on new initiatives and are regularly recognized for their inventiveness and rewarded in material and nonmaterial terms for the success of their work. Creating a climate of collaboration is a time-consuming activity, but once a campus culture is infused with that sense, the impact can be exponential. In turn, institutionalizing the effort can reinforce the bridging of business and academic affairs. Having representation from both operational areas, endorsing co-leadership, and broadening the representation to include members of the nearby community all contribute to building this culture of entrepreneurship. The university has benefitted from the creation of the Council on the Environment (COTE), which now has a long history of facilitating such conversation and which has, in fact, built its monthly agenda around a very simple model of *celebrating, facilitating,* and *anticipating*—a neutral-enough framework to encourage all parties at the table to contribute to the good of the institution, its surrounding community, and the larger social, economic, and environmental landscape (see table 8.3).

To understand more fully the importance of COTE, it will be helpful to briefly review its history and accomplishments. Recommendations for the creation of the council came from the work of the Green 2 Committee, which was charged with examining how the university could implement the tenets of the Talloires Declaration, which had been signed in 1999. The Green 2 Committee recommended numerous action items that were content specific but recognized the need to put in place an operational structure that could assure the long-term maintenance and oversight of the tasks of implementing the Talloires tenets. In fact, the tenth tenet of that declaration requires that a secretariat be established to carry out the work. The Green 2 Committee recommended that a council on the environment be created whose membership would represent the full breadth of the university and surrounding community and that day-to-day administrative support would be provided by an existing unit on campus that was not within a given academic college. The Center for Energy Research/Education/Service (CERES) was asked to serve as that administrative home.

COTE has now been in existence now for some eight years; it meets on a monthly basis, and a key initial example of its impact on university life was its recommendation that Ball State adopt a public Sustainability Statement—drafted by the council, shared with the university senate, academic deans, senior staff, and the board of trustees, all of whom endorsed its language and its placement on the university's web page. Many of the recommendations of the council have been policy related, including the request that the university adopt LEED silver certification as a standard for all new construction and building remodeling. This was then woven into our strategic plan. COTE has also supported the removal of mercury from the campus, endorsed a smoke-free-campus policy, the use of post-consumer recycled content paper stocks, printer cartridge recycling, acquisition of hybrid electric vehicles for the university

Table 8.3. Ball State's Council on the Environment (COTE)

COTE provides leadership for initiatives at Ball State University and in the surrounding community that promote the sustainable use of natural resources and the protection of ecological systems that sustain life.

To encourage the development of an environmentally responsible campus community, the Council on the Environment will:

1. Provide a forum for the open exchange of information pertaining to environmental issues;
2. Increase awareness of the environmental and social implications of university operations;
3. Promote a strong, positive environmental ethic that motivates sustainable practices in all university activities;
4. Promote communication and education about sustainability issues and achievements;
5. Promote outreach and educational activities that encourage sustainable practices in the broader community.

Three Emerging Trends

- Asynchronous online courses and degree offerings as a "green" strategy
- Proliferation of rating and ranking systems that use widely differing metrics
- Emphasis on closing the financial circuits: addressing returns on investment, investment responsibility, and the erasure of split incentives

Three Best Practices

- Encouraging a multifaceted, campus-wide engagement on sustainability issues
- Formalizing empowerment through decentralized, unit-level sustainability planning
- Hosting a monthly committee or council on the environment

fleet, and creation of an awards system for campus and community recognition. While some calls for action have been modest, others have been substantial, but all have been essential to the social, environmental, and economic health of the campus community.

For example, COTE's recommendation for the university to move to the use of hybrid-electric vehicles in its fleet initially produced significant push-back; members were told that a cost-benefit analysis of such vehicle adoption would be required before a decision to implement was possible. The council simply asked for a copy of the cost-benefit template used to make purchasing decisions for the more conventional gasoline-powered automobiles and were told no such cost-benefit analysis was made. (As is often the case, initial resistance is predicated on a non-level playing field of decision criteria.) Having exposed this rather embarrassing reality—and with test drives by selected university administrators—the decision to support the COTE recommendation was an appropriate next step for the institution. BSU acquired an initial group of these vehicles for its fleet. Perhaps even more significantly for chief financial officers and sustainability directors in particular, these vehicles have been retained beyond the normal several-year cycling of replacement and are intentionally being pushed to yield more understanding of their resilience and serviceability. Fleet management administrators have since acquired additional vehicles of a different brand to fill out the inventory, a case in point of a good idea catching fire and having a life of its own.

Perhaps one of the more interesting stories involves the impact of COTE on the surrounding community. Minnetrista Cultural Center, a local philanthropically supported facility for community programs, has chosen to take on the BSU model and has woven sustainability principles into its strategic planning while implementing empirical study of green-roof technology at its own facility. COTE tried for years to bring green roofs into the campus construction palette and continues to encounter significant resistance. In this instance, the community is leading the way with its own empirical testing and evaluation of this technology. Members of COTE take the view that with a successful completion of the study and with the anticipated defensible economic and technical findings, the university is likely to achieve a comfort level for its own adoption.

As mentioned above, the council adopted the practice of giving annual awards to on-campus and off-campus community members for Green Initiatives, Exemplary Practices, and Lifetime Achievement. The awards bestowed to date acknowledge student enterprise, individuals and academic units evidencing best practices, and community members involved in everything from drug recycling to animal rescue. In fact, after almost a decade of institutional service and multiple programs that have transformed university life, the council's work has set the stage for the staff of business affairs and academic affairs to achieve a

Single Most Important Piece of Advice for a Young Leader Entering the Profession

- Awaiting the perfect decision is a form of procrastination; acknowledge that there is no single answer and that things will be messy but that action cannot wait.

A Decision One Would Make Differently with Hindsight

- Approaching sustainability as an opportunity for financial planning and leveraging and, more specifically, discovering the significant return to be achieved through the installation of a major geothermal district heating and cooling system flexible comfort level with collaboration on many issues extending beyond sustainability. Business affairs administrators routinely approach faculty members to assess collaboratively the role of research and education in facilities funding and management.

A major example of this is the new geothermal district heating and cooling system approved just this year by our board of trustees. Embracing environmentally sensitive initiatives is new territory for business affairs; likewise, making use of campus facilities as a teaching and research subject is new territory for many in academic affairs. For this project, the business affairs office has decided to consider an alternative to Ball State's coal-fired heat plant—a bold departure, considering Indiana's history as a coal-producing state. Still, the university is moving ahead with the project as a multiyear commitment to its physical buildout and a cutting-edge research opportunity that looks to enhance the marketability and return on investment of this technology. Successful operation and meaningful research discovery will depend on a healthy continuing collaboration of both houses.

Tied to this collaboration, of course, is the continued search for funding to supplement the state-level appropriations that enable the first steps in this transformation from a coal-fired heating facility to a geothermal heating and cooling system driven by electrical power. As strategies unfold, the university will pursue additional sources of green power production both onsite and over the grid. We see this as an opportunity for supporting and encouraging the growth and development of our green energy economy. Similarly, the funding question is increasingly a challenge for university leaders attempting to accomplish major changes on their campuses. Shadowbox 2 at the end of this chapter lists the kinds of funding mechanisms used throughout campuses in North America to source new lines of support for both small- and large-scale intervention, but beyond the question of money itself, presidential leadership in collaboration with community engagement has made the critical difference at institutions like Ball State.

Put differently, leveraging leadership both "at the top" and "in the trenches" is a simultaneous and necessary enterprise. For presidents and trustees, clear communication, transparent information flow, and the championing of the holistic nature of the sustainability issues we face today are essential, and this must be complemented by the work and commitment of faculty, staff, and students.

Using Multi-Modal Leadership Techniques

In accepting a presidency, one assumes a multitasking life. Certainly, this reflects the complexity of the home institution and the numerous

A Sustainability Myth

- Sustainability initiatives always cost an institution more money.

local, state, regional, and even national networks of connectivity to the social, environmental, and economic arenas. The leadership challenge is to strike a balance: being "reactive," when necessary, to the complexity and pressures of day-to-day operations, while also proactively leading in the development of a long-range vision for the university. In response to these challenges, and beyond the conventional descriptions of how presidents must be able to be in three places at once, we offer in what follows a somewhat different model of presidential leadership that is working at Ball State, as the university moves toward a broader and deeper embrace of the principles of sustainability: multi-modal leadership.

In principle, multi-modal leadership requires attention to differing scales of timeline for action and differing conventions of administrative practice. More specifically, some opportunities arise that benefit from *swift authoritative moves* on the part of the president. At Ball State, the most recent example of this was to secure the support of the board of trustees to permit the university to install the aforementioned district-scale geothermal heating and cooling system. While it took some time to do the background research and to develop proofs of concept to assure the administrative cabinet and board members that it would be a cost-effective and technically sound project, once we had the data in hand, the decision to move forward was made quickly. A related example was the decision to become a founding member of the Leadership Circle of ACUPCC signatories and to employ the president's bully pulpit to state a commitment by the institution. Such opportunities sometimes present themselves in unpredictable ways, but taking immediate action with confidence is only possible when understood in service to the longer-term vision of the university.

In the area of strategic planning, colleges and universities typically appoint task forces charged with bringing forward recommendations; the faculty as a whole, as well as administrators, professional staff, and student representatives participate in shaping these recommendations. Many posit that the *process* of strategic planning is more important than the specific goals and objectives set forth in a final written document. While this is largely true, the conventions of strategic planning also offer a window of opportunity to introduce sustainability.

Funding Mechanisms for Campus Sustainability

Administrative Allocation

Providing dedicated funds is a form of investment for which a percentage return can be anticipated. In the case of the Harvard Green Campus Initiative, the dedicated allocation functions as a revolving loan fund. It is possible to set these allocations up to yield, and that residual can pay back the original stipend and thus allow the fund to become self-perpetuating.

Alumni Sustainability Fund

Appealing to a sizeable alumni base of a major university can yield substantial aggregate impact if requests are made for modest levels of contribution. This is in some ways more meaningful than securing a single major donor because of the breadth of connectivity that is established by the contribution.

Chargebacks

Although many universities are centrally budgeted and utility costs are tracked through a single gate, the sub-metering and apportionment of operational expenses to the respective administrative units on campus can erase the challenge of the split incentive. If a given department or college saves operational money on utilities and can reallocate those saved dollars to other purposes, the incentive will be there at a grassroots level to accelerate the needed transformation.

Class Gifts

Certainly any graduating class may find it desirable to invest in elements of ongoing campus sustainability. Rather than funding a new gateway or sign for the campus grounds, the purchase of building-integrated photovoltaic panels or similar technologies offer significant return on investment and maintain the intent of the long-term presence of the class gift as part of the campus physical facility.

Endowments

The holdings of universities can be treated in at least two ways. In one sense, sustainability offers a marketing opportunity for garnering additional endowment support, but secondly and more importantly, the decision to invest in social, economic, and environmental equity-based instruments is not only the "right thing to do" but often yields a higher return on investment than conventional stock and bond holdings. This is especially true if some of those endowment monies are used to establish a revolving loan fund, as mentioned in the Harvard example.

Fee-for-Service

Decentralizing and articulating the overhead costs of "non-green" vs. "green" supplies and expense budgeting at the unit level can offer substantial incentive for the needed change. This comprises another form of chargeback that can be readily accounted for using bar code and computer tracking techniques.

Incentives

The American Recovery and Reinvestment Act (ARRA) offers significant new funding for transitioning the country to a green economy. This links well to the research, education, and service opportunities in the sustainability arena. Other incentive programs include tax credits for energy conservation design that can be "given back" to the design firm as part of the compensation package for their work.

Parking Fees

Charging differential prices for parking near buildings is common on most campuses; further differentiating this practice with preferred access for high-mileage vehicles is an opportunity easily implemented.

Payroll Deduction

Small incremental dedication of funds from the full faculty and staff of an institution can aggregate substantially in a single fiscal year. Although such an internal "tax" is not often used except in the case of donations to public agencies or set-asides of money for development funds, this is a relatively painless way of invoking participation by the full university community (see Student Fees below).

Performance Contracting

Many institutions are making use of the Clinton Climate Initiative funding in which energy service corporations are providing the front-end capitalization of energy conservation retrofits to campus buildings, using contract language that returns a percentage of the resulting energy savings to the energy service company as its return on investment. Typically, these contracts have a bracketed time in which such annual payback is made, after which the residual savings accrue solely to the institution.

Foundations

More and more foundations are shifting their emphasis to the dimensions of sustainability. There is no shortcut to leveraging this opportunity, but the good news is that the shift is occurring.

(*cont'd*)

Funding Mechanisms for Campus Sustainability

Revolving Loan Funds

As mentioned above, in some instances institutions have used endowment monies to establish such funds and replenished the funding base with the savings resulting from energy conservation measures. An important consideration is the bundling of strategies, so that investments that give a lower percentage return can be mixed with those that yield a higher percentage to garner an average net performance that is acceptable; this assures the completion of projects that would have been set aside based on strict analysis of the single return-on-investment percentage.

Savings from Sustainability Measures

Savings from sustainability measures are typically rolled back into the revolving loan fund if they were used as the form of capitalization of the energy or material conservation initiative. However, once the capital holdings of an institution have reached reasonable levels of performance, the long-term potential of this technique is to turn these savings to other programs.

Student Fees

In most instances, universities that have adopted green fees paid by students have done so in response to student initiative. Students will typically mount a campus campaign to convince their peer groups that this is a sensible move and that for a modest contribution each semester, the aggregation of money can be significant. The challenge is for the student groups to negotiate an agreement with the university on where to invest the resource. Thus, much front-end planning is needed and coordination is required before the campaign is mounted and the fees are accrued.

The Ball State community chose to use its most recent strategic planning process to call for unit-level sustainability planning on the part of all administrative units of the university. In the process, chief planners received more than one hundred such unit-level plans, each organized according to the common framework used in the Sustainability Tracking Assessment and Rating System being promoted by AASHE. The senior leadership team's embedded goal was to prime the participation of faculty and staff in understanding STARS as a new national model to assess multi-levels of university performance and to get out in front of this development with first drafts of sustainability plans that reach deep into the administrative structure of our institutional community. Perhaps not surprisingly, some administrative units accepted this task with a matter-of-fact response, while others used the call as an opportunity to "grow the conversation" within the departmental ranks. Ball State's Honors College actually asked its students in consultation with the faculty to develop the unit-level sustainability plan for the college.

At the same time, not all presidential decisions are swift, and not all plans run smooth and straight. Much of the work of the presidency is necessarily long term and requires striking a vision of where the institution needs to be by some future date, especially in the face of climate action planning wherein we must set targets for achieving climate neutrality. In this context, routine communication and day-to-day referencing from a leadership position is necessary, particularly in the finance and budgeting considerations of the institution when managing trends in state-level assistance, market pressures on tuition and fees, and the tightening of philanthropic support.

A key example of the value of careful community engagement and consensus-building was Ball State's decision to become a smoke-free campus. This process started with a statement from the office of the president that the university was going to evaluate becoming a smoke-free campus. That statement was also included in the university's strategic plan. The process was bolstered by an immersive learning project initiated by our faculty that started the conversation on campus. That learning project provided pro and con arguments and led to several student demonstrations related to going smoke free. Subsequently, the Professional Staff Council was asked to express their thoughts on this issue. This was followed by a student survey and polling of the departments by the University Senate, and the proposal was formally endorsed by COTE resolution. The process ended with a pronouncement that there was enough consensus on campus to go smoke free—a presidential determination—followed by a notice period of four months before enforcement began. The whole process took two years, perhaps longer than originally envisioned, but it enabled the entire campus to get used to the idea and to realize that the train was moving down the track.

Equally important are those cycles of events and ceremonial activity on a campus that present venues for reminding the university community of its vision and its progress. At Ball State, we have found it helpful to have hosted, since 1996, the Greening of the Campus Conference series wherein, on a biennial cycle, we bring in keynote speakers, call for academic paper submissions, and provide peer review and publication in conference proceedings. While the university community learns much from those who visit our campus and share advances made at their home institutions, perhaps more importantly, in retrospect, the university has used the rhythmic occurrence of this biennial gathering as a way to remind our own university community of our positioning on the national and international stage and of our progress in achieving our sustainability goals.

These examples point to the need for a president to understand fully the culture and practices of the institution he or she leads. As a new president, one must take time to come to terms with internal and external institutional memory and to survey and inventory past success stories, all the while looking for ways to align such experience with developments on the frontier of sustainability leadership. As a continuing president, one must get outside the habits of mind employed for day-to-day responsibilities and recognize that the deep knowledge of one's institution will best be put to use in visioning the role of higher education as a shepherd of social, environmental, and economic equity.

In the end, presidents are administrators as well as leaders, and it is incumbent on them to ensure that all parties needed at the table show up and that their contributions are given ample space and time for impact. This reflects again the earlier observations that one must be operating in top-down *and* bottom-up fashion at all times. Certainly, connecting the dots is everyone's territory.

Socializing the Board of Trustees

Trustees, to be effective, need steady updates on all key issues, and sustainability initiatives are no exception. In fact, thinking seriously about sustainability may be a new assignment for board members on many campuses. Whatever trustee experience levels are, initial steps should include individual briefing sessions with new members to acquaint them with practices, protocols, and policies of board operation and to educate them about the history of the college or university and recent board actions. In a larger sense, the challenge is to illustrate to the board as a whole the opportunities presented to the institution when it recognizes the *interdependence* of the social, environmental, and economic dimensions of sustainability-driven thinking.

Within any board there will arise champions in particular topical areas, and savvy presidents and chairs are increasingly seeking new trustees with experience in sustainability planning and projects. In fact, issues related to sustainability present a broad-based opportunity to make effective communications among board members and then between the board and the leadership team. However, since sustainability in principle engages so many dimensions of our operational world, it can seem especially daunting to a new board member. The challenge in all cases, however, once new board members have been brought up to speed and continuing board members have acknowledged their place at the table, is to be quite deliberate in the mechanisms by which information is shared and the membership of the board is kept informed.

Fundamental to that effort is assuring a consistent, repeated information flow to and from the board—meeting on a monthly basis, with intermittent correspondence through email and phone conversations. A predictable pattern of operation for board meetings is equally important. At Ball State University, we have found it useful to offer program presentations as features within the board sessions each month, using the expertise of our faculty to communicate their research, education, and service work and its impact on the regional and statewide economy, citizenry, and natural resource base. Consistency of information flow is driven in part by clear themes, such as the branding of the educational experience at Ball State as an "Education Redefined" or, as related in this chapter, the branding of Ball State as a "green university." And while such singular terms can offer identity and moorings for conversation, one must go beyond such simplification to the fully vetted content implied therein. Thus, we host the program presentations during the board meetings and work diligently to script communiqués from the office of the president to the entire university community. Finally, we work intentionally with the offices of enrollment management and corporate communication to ensure that members of the institution are fully informed about the progress we are making in meeting the goals and objectives of our current strategic plan and the role of the board of trustees.

Connecting to other private and public institutions in the region is another important role for the presidency in service to the sustainability agenda. In many instances an institutional partner will be considerably further along in its development of sustainability programs, policies, and action items. For example, Ball State's Council on the Environment (COTE) is one of the longest-running groups of its kind, and planners from colleague institutions have regularly queried Ball State leaders on its formation and effectiveness. In other cases, we have much to learn from notable activities by a sometimes smaller college or university. Goshen College, for example, executed remarkable leadership with the design and construction of its Merry Lea Environmental Center—the first LEED platinum–certified building in the state of Indiana.

Conclusion

This chapter has outlined five starting points for presidents and trustees who are guiding their college or university toward sustainable goals. At Ball State, this work has been found to be most effective when it is undertaken through actions that are simultaneously bottom-up and top-down. In this process, faculty members and administrators at the university have learned that sustainability should not be considered an exotic specialty to be brought *over to* all that we do as agents of higher education; rather, sustainability should be sought, and found, as integral to and *within* the multiple layers of educational policy and action on our campuses, including research, operations, administration and finance, community relations and partnerships, diversity, access and affordability, and human resources.

An advantage of beginning this journey now is the wealth of resources currently available to higher education leaders. Not only is this larger volume a key resource, but the numerous organizations cited in this chapter and the many resources published on the various acronym web pages all comprise the tools and information needed to position a school in the twenty-first century. Certainly, initiatives like the ACUPCC and STARS have brought sustainability into much brighter light nationally, and presidents now have the advantage of annual reporting from such groups as the Sustainable Endowments Institute and the AASHE Digest, among others. In sum, presidential leadership for sustainability objectives must reflect the understanding that we are uniquely positioned—individually and collectively, as part of an international population—with the explicit task to shape the future for many citizens and, only slightly less importantly, to make that future more rather than less sustainable. For higher education leaders, the moment remains filled with possibility, risk taking, and a need for innovation.

APPENDIX: REGIONAL SUSTAINABILITY ORGANIZATIONS IN HIGHER EDUCATION

West

Alaska, Arizona, California, Colorado, Hawaii, Idaho, Montana, Nevada, New Mexico, Oregon, Utah, Washington, Wyoming

Washington Center for Improving the Quality of Undergraduate Education
www.evergreen.edu/washcenter/

Midwest

Illinois, Indiana, Iowa, Kansas, Michigan, Minnesota, Missouri, Nebraska, North Dakota, Ohio, South Dakota, Wisconsin

Midwest Regional Collaborative for Sustainability Education
www.mrcse.org/

Michigan State University Office of Campus Sustainability
www.ecofoot.msu.edu

Indiana Consortium for Education towards Sustainability (ICES)
www.ices.ws/

Upper Midwest Association for Campus Sustainability (UMACS)
www.umacs.org

Northeast

Connecticut, Maine, Massachusetts, New Brunswick, Newfoundland, New Hampshire, New Jersey, New York, Nova Scotia, Ontario, Pennsylvania, Quebec, Rhode Island, Vermont

Green Campus Consortium of Maine
www.megreencampus.com/

Northeast Campus Sustainability Coalition (NECSC)
http://sustainability.yale.edu/

South

Alabama, Arkansas, Delaware, Florida, Georgia, Kentucky, Louisiana, Maryland, Mississippi, North Carolina, Oklahoma, South Carolina, Tennessee, Texas, Virginia, Washington, D.C., West Virginia

South Carolina Sustainable Universities Initiative (SUI)
www.sc.edu/sustainableu

Associated Colleges of the South (ACS)—Environmental Initiative
www.colleges.org/enviro/

SOURCE: *From* www.umacs.org/sustainability-organizations-higher-education

CHAPTER NINE

Not So Fast: A Dose of Reality about Sustainability

Thomas Buchanan and Tara Evans

The Making of a Contrarian View

Sustainability has become a nondescript workhorse concept that can be harnessed to pull just about any commercial wagon. "Sustainability" and "green" have become parts of pop culture, and although they are not synonymous, people tend to view them as such. For more than a decade, companies have been using green virtues to sell everything from clothing, cars, and appliances to toilet paper, and this approach has persuaded customers. In 2006, the National Marketing Institute calculated the Lifestyles of Health and Sustainability Market at $206 billion.[1] In 2007, Walmart pledged to sell one hundred million compact fluorescent light bulbs (CFLs).[2] Both were marketing ploys based on sustainability. Yet what does sustainability really mean? For some people, it means turning off the light bulb. For others, it means buying an electric car. This chapter emphasizes a contrarian view. Sustainability has to mean something vastly different from tray-free cafeterias and student-run gardens, and it has to have much deeper dimensions, or our world will not be sustainable in 2025, much less in 2050.

The sustainability movement in higher education is often just a discussion about populist notions that are undefined at best and border on being an unfortunate deflection of real research on global problems. These notions are being picked up by marketing firms because colleges and universities sense that sustainability is attractive to potential college students. However, what is needed now is serious and focused discussion about policies that will change global behavior. To facilitate them, higher education institutions, and in particular their presidents, boards, and chief academic officers, need to address both sides of the sustainability equation: the demand side and the supply side. These discussions must address changing core values, and the planning for them must have more depth and breadth than many of the current sustainability conversations being held at well-meaning institutions. Campus communities need a dose of reality about sustainability to plan effectively for our ecological and economic future.

Readjusting the Sustainability Equation

The first step in broadening sustainability discussions is to readjust our understanding of the equation. The current discussions are heavily weighted on the demand side of the model. Sustainability discussions, however, will never achieve full potential unless we recognize the other side of the equation—*supply-side sustainability.* Here is an example: Higher education institutions are substantial contributors of greenhouse gas emissions. As such, the American College & University Presidents' Climate Commitment (ACUPCC) was formed to encourage campuses to formulate a plan for carbon neutrality.[3] The University of Wyoming signed on in August 2007. Part of the commitment entailed a comprehensive inventory of all the university's greenhouse gas emissions. The results were interesting, to say the least. In 2008, approximately 46 percent of the University's greenhouse gas emissions were attributable to purchased electricity, 39 percent to on-campus stationary sources, 12 percent to transportation, 1 percent to agriculture, 1 percent to solid waste, and 1 percent to refrigerants and chemicals. Of the emissions attributable to on-campus stationary sources, approximately 90 percent were from coal and 10 percent were from natural gas.[4]

Three Emerging Trends

- Recognizing not only the demand side of the sustainability equation but also the supply side by using a combination of everything we have—conservation, renewables, efficiency, gas-fired power plants—and using it wisely and efficiently.
- Reexamining our scale. Too much of the sustainability discussion is focused on too small a scale. Our challenges are global; so too will be our solutions.
- Broadening the discussions that are occurring on campuses to teach students the real magnitude of the problem, not just how to be ecologically minded.

Three Best Practices

- Partner with governmental institutions and the private sector to address the supply-side energy technologies and resources that will play a role in the transition to a sustainable energy future.
- Play a vital role in our energy future by producing the technologists, scientists, engineers, lawmakers, and policy makers who will lead us in sustainability.
- Develop a long-range development plan for the campus, beginning with a four-pronged approach: efficiency and energy conservation, renewable energy supply strategies, resources management, and alternative transportation.

Wyoming derives a sizable portion of its revenue from mineral development. Indeed, a record volume of coal—in excess of 462 million tons—was mined in Wyoming in 2008.[5] Since Wyoming is a supply-side state, with an abundance of easily accessible coal, the University of Wyoming will not be reducing its reliance on coal in the foreseeable future. According to the National Coal Council, fossil fuels provide approximately 85 percent of the world's energy.[6] In 2030 that figure will still be about 85 percent. Of that 85 percent, coal makes up 28 percent and is irreplaceable as the fuel of the future because it has abundant supply, availability, versatility, affordability, and emerging receptivity to carbon capture. Coal generates more than half of our electricity and is still the most plentiful fuel source in the United States, projected to last more than two hundred years at today's level of use.[7] Thus, coal must remain a part of Wyoming's and the nation's fuel mix to avoid potentially devastating economic consequences.

Despite Wyoming's and the university's reliance on and commitment to coal, the University of Wyoming can still be a leader in supply-side sustainability. Demand-side solutions will never be long-term solutions without the supply-side energy technologies and resources that will play a role in the transition to a sustainable energy future. The ability to feed our world energy needs over the next fifty years will require maximizing all forms of energy. As Thomas F. Farrell II, chairman, president, and chief executive officer of Dominion Energy, reminds us, "You can shut down all the coal plants in the United States, and if you only replace them with gas-fired power plants it puts you where we are today by 2050, given economic and population growth."[8] The key is using a combination of everything we have—conservation, renewables, efficiency, gas-fired power plants—and using it both wisely and efficiently. College and university leadership teams need to stop pretending that renewables are the end-all fix.

Just to be clear, the demand side is a critical part of the equation. We should not minimize demand-side fixes like hybrid cars, fluorescent light bulbs, and recycling. But the world's energy woes will not be solved by only modifying the demand, such as driving a car that gets forty miles per gallon as opposed to one that gets twenty. Supply-side sustainability will only come about through changes in policy, and changes in policy are about changes in belief, which can make those experiencing the change very nervous. Changing a nation's worldview is an exasperatingly slow and difficult process. One difficulty with the current sustainability movement is that it has allowed itself to become too narrowly defined. Michael Shellenberger and Ted Nordhaus, career environmental activists, note that sustainability has become a special interest, which has diminished the importance and momentum that the movement needs.[9] One solution to the problem of special interest politics is to form effective alliances. Going forward, laborers, conservationists, and industrialists need to work together; and presidents and provosts, in particular, need to frame the discussions on their campuses.

Collaboration: The Way Forward for Colleges and Universities

Colleges and universities also need to partner with other governmental institutions and the private sector. As a case in point, Wyoming is investing millions in the University of Wyoming to develop expertise in energy resources. The Clean Coal Technologies Research Program was enacted by the Wyoming legislature in 2007 to stimulate research that will enhance and improve clean coal technologies and emphasize the use of sub-bituminous coal at high altitudes. Research proposals are solicited each year from academic institutions and private industry on how to improve the use of Wyoming's coal resource. The University of Wyoming has also recently embarked on two important partnerships—one with the National Center for Atmospheric Research (NCAR) and one with GE Energy. The university has partnered with NCAR to locate its new supercomputing facilities in Wyoming, allowing UW scientists access to tremendous computing power that could improve analysis of the state's complex geology and enhance understanding of the state's large-scale carbon sequestration capabilities. GE Energy is also working with the university to develop an advanced gasification research and technology center. The center will consist of a small-scale gasification system that will allow the University of Wyoming and GE researchers to develop advanced coal gasification technology solutions for Powder River Basin and other Wyoming coal.

Another significant partnership at the University of Wyoming involves carbon sequestration. In the United States, per capita carbon emissions equal an estimated 20.6 tons per year.[10] In Wyoming, per capita carbon emissions equal an estimated 127 tons per year—more than six times the amount of carbon as the rest of the nation. This is because Wyoming residents are "credited" with carbon emissions created during the production of electricity that is transmitted out-of-state for people in other states to heat their homes and run their air conditioners—all the while pretending they are carbon neutral. Similarly, as the nation's largest "net" exporter of energy, Wyoming's emissions per capita are number one in the United States. This fact has not gone unnoticed. In March 2008, Wyoming became the first state to adopt comprehensive geologic carbon sequestration legislation. This legislation essentially functions as a first step toward a statutory and regulatory framework surrounding the capture, transportation, siting, operation, and closure of carbon.[11] To understand further the potential to provide long-term storage of carbon dioxide, the University of Wyoming, the state of Wyoming, the federal government, and the private sector are teaming up to research carbon sequestration opportunities in Wyoming. Principal investigators from across the UW's Colleges of Arts and Sciences, Agriculture and Natural Resources, and Engineering and Applied Science, as well as personnel from the Wyoming Geographic Information Science Center, the Wyoming State Geological Survey, and industry giants Baker Hughes Incorporated and ExxonMobil, are evaluating the potential for carbon sequestration opportunities in one of the deep saline aquifers of the Moxa Arch and Rocks Springs Uplifts, a capacious geological structure in southwestern Wyoming. This effort is being funded by an award from the U.S. Department of Energy, a match from the University of Wyoming, a state appropriation, and private industry.

Single Most Important Piece of Advice for a New Professional

- "Sustainability" and "green" have become part of pop culture. Despite the current sustainability fad, colleges and universities can and must step up. No institutions are more equipped to take a leadership role than those in higher education.

A Sustainability Myth

- Colleges and universities believe that by hiring a vice president or director of sustainability, they are addressing the issue. Instead, higher education institutions should be sponsoring global discussions.

Importantly, all of these energy-related partnerships and alliances are integrally related to the University of Wyoming's core academic mission and are coordinated in part by the university's recently established School of Energy Resources (SER). This school facilitates interdisciplinary academic and research programs in engineering and science, economic, and environment and natural resources policy to address critical energy-related issues faced by our global society. SER director Mark Northam emphasizes that the establishment of this school was an important step in moving toward sustainability at the university and in Wyoming: "The state is blessed with abundant energy resources and environmental treasures. Our challenge is to produce and consume energy resources while minimizing and mitigating environmental impacts. SER works with University colleges and the Haub School and Ruckelshaus Institute of Environment and Natural Resources to develop technologies and provide for an energy future that is sustainable in the long-term."[12] Collaboration among higher education institutions, laborers, conservationists, industrialists, policy makers, and many others is a crucial step in addressing the supply side of the sustainability equation. These partnerships will also broaden sustainability discussions on campuses and give presidents and trustees some excellent opportunities to become more engaged in these issues over the coming decade.

Changing Our Core Values

The stereotypical American defines the good life as owning a bigger car and making more money. Americans are not hardwired to scale back—we are energy driven and resource consumptive. Food writer Michael Pollan emphasizes this point, defining the environmental crisis as a crisis of character: "It's really about how we live. The thought that we can swap out the fuel we're putting in our cars to ethanol, and swap out the electricity to nuclear and everything else can stay the same, I think, is really a pipe dream."[13] At the 2008 Senior Executive Energy Summit held in Jackson Hole, Wyoming, topics included domestic energy policy, energy access, and climate change. Summit panelists noted that by 2030, the world is projected to increase its energy consumption by 50 percent.[14] It will cost $22 trillion to meet this demand. According to Shellenberger and Nordhaus, the challenge is of monumental size and complexity—remaking the global economy in ways that will transform the lives of six billion people.[15] One of the only factors that will modify our consumption of energy resources, however, is cost. We will continue to use fossil fuels until they become so expensive that we cannot pay the costs. That is our nature. Our country is built on cheap energy, and only when the prices begin to directly affect consumers and their quality of life will we start to modify our consumption.

Take natural resources, for example. Daniel Kammen, a distinguished professor of energy at the University of California–Berkeley, identifies the challenges ahead for developing renewable energy: "Suppliers of renewable energy must overcome several technological, economic and political hurdles to rival the market share of the fossil-fuel providers. To compete with coal-fired power plants, for example, the prices of solar cells must continue to fall. The developers of wind farms must tackle environmental concerns and local opposition."[16] If we take into account the world's competing energy resources, renewables produce only a fraction of the global electricity generation. Coal, oil, natural gas, nuclear, and hydropower make up 98 percent. Only 2 percent of global electricity is generated by non-hydropower renewables like biomass, geothermal, wind, and solar. Kammen calculates that the electricity produced by solar cells has a total cost of twenty to twenty-five cents per kilowatt-hour. Compare this to the four to six cents for coal-fired electricity, five to seven cents for power produced by burning natural gas, six to nine cents for biomass power plants, and an estimated two to twelve cents for nuclear power. The difference is staggering.[17]

Renewable resources are a key resource for our future, but they are currently far from cost effective. That means that most of the American public will not be interested. The public needs to be better informed on the realities of economics and regulations. The solution is a coherent national and international energy policy that centers on energy security, intelligent decision making, and government backing. In 2006, the Wyoming legislature provided funding and authorization for the development of the School of Energy Resources (SER) at the University of Wyoming.[18] SER's objectives are to provide nationally competitive undergraduate and graduate instruction in energy-related disciplines; to advance Wyoming's energy-related science, technology, and economics research; and to support scientific and engineering outreach through dissemination of information to Wyoming's energy industries, companies, community colleges, and government agencies.[19] SER is committed to energy education that influences the broadest possible student cross-section; as such, the Wyoming legislature allocated new positions in 2006 for energy teaching and research at the University of Wyoming. Additionally, an Energy Summer Institute at the university helps foster an environment in which to develop strong and lasting links with Wyoming's K–12 teachers, counselors, and students. The institute is designed to provide participants with exposure to the challenging energy problems facing the world and the exciting solutions on the horizon. Courses offered by the institute include powering the future with renewable wind energy, energy-efficient architecture, and an introduction to global position systems, geographic information systems, and remote sensing technologies. Importantly, the matching grant fund program allows the university to garner significant external funds to meet faculty and academic professional research objectives. These academic, research, and outreach programs have demonstrated SER's and the university's leadership role in the energy equation.

Also significant to the University of Wyoming's leadership role are the relationships, on and off campus, that SER cultivates. Among these important collaborations is the substantial ongoing interaction SER has with the Haub School and Ruckelshaus Institute of Environment and Natural Resources at the University of Wyoming. The Haub School has a tradition of active involvement in a variety of energy-related issues, and this partnership encourages policies and directions that advance both the mission of SER and the Haub School. The university has also added a new major to its repertoire: Energy Resource Science. This bachelor of science degree program involves a collaboration between SER and the university's Colleges of Arts and Sciences, Engineering and Applied Science, Agriculture and Natural Resources, Business, Education, and Law, as well as the university's Haub School. The goal of the Energy Resource Science major is to offer a diverse curriculum that combines engineering, science, business, law, and natural resources in order to build a fundamental understanding of interactions and trade-offs between energy, environment, policy, and the economy.

Higher education institutions have the human capital and the expertise to be leaders in the energy equation. We can educate students and the public on the broad spectrum of energy and demand. We can also provide focused, sustained research and development. According to Mike Chesser, chairman of the Electric Power Research Institute, "The great challenge of our time is to operate our existing electric infrastructure while developing new technologies to achieve unprecedented levels of efficiency and sustainability."[20] Both colleges and universities have the capability to produce the technologists, scientists, and engineers that will lead us in achieving sustainability goals. We can produce an energy-educated workforce who will eventually become the lawmakers and policy makers who will help us begin changing our core values.

Broadening Campus Discussions: The Challenges Ahead for Presidents and Leadership Teams

At a time when the world is struggling to define a vision for the future of energy and the economy, the conversations occurring on campuses are too often one-dimensional. We need to ask more of our presidents, chief academic officers, and trustees in particular. Michael Maniates, a professor of political and environmental science at Allegheny College, talks about the challenge of individualizing responsibility. The American response to the current environmental crisis, says Maniates, is to blame degradation on individual shortcomings: "It embraces the notion that knotty issues of consumption, consumerism, power and responsibility can be resolved neatly and cleanly through enlightened, uncoordinated consumer choice."[21] This individualization, however, prevents us from "collectively changing the distribution of power and influence in society—to, in other words, 'think institutionally.'"[22] That, unfortunately, is what is now happening at many college and university campuses. We are thinking individually, rather than institutionally. Groups of well-intentioned students and faculty assume that if they alter their lifestyle others will follow. This is an excellent notion, but it is inadequate. We must teach students the real magnitude of the problem, not just how to be ecologically minded. This approach means that we must deepen and broaden our outlook on sustainability beyond many current conversations. Shellenberger and Nordhaus argue that leaders are failing to articulate a vision of the future equal to the magnitude of the crisis.[23] We tend to focus an excessive amount of attention to policy fixes like pollution control and higher mileage standards. Maniates provides some striking examples:

> We are individualizing responsibility when we agonize over the "paper or plastic" choice at the checkout counter, knowing somehow that neither is right, given larger institutions and social structures. We think aloud with the neighbor over the back fence about whether we should buy the new Honda or Toyota hybrid-engine automobile now or wait a few years until they work the kinks out. . . . [W]e ponder the "energy sticker" on the ultra-efficient appliances at Sears, we diligently compost out kitchen waste, we try to ignore the high initial cost and buy a few compact-fluorescent light bulbs. We read spirited reports in the *New York Times Magazine* on the pros and cons of recycling while sipping our coffee, study carefully the merits of this

> and that environmental group so as to properly decide upon the destination of our small annual donation, and meticulously sort out recyclables.[24]

Presidents, provosts, and faculty leaders need to move their campus conversations beyond tray-free cafeterias, recycling, and student-run gardens to a more penetrating look at what sustainability thinking could look like ten, twenty, even fifty years in the future, rather than simply during the present budget cycle.

One key to being a leader in sustainable policy development is long-term campus planning. In the *New England Journal of Higher Education,* Greg Havens, Perry Chapman, and Bryan Irwin, architects from Sasaki Associates, reported on the University of Maine's approach to sustainability in campus planning. Land grant universities like the University of Maine can be emerging leaders in the sustainability movement based on their deep historic ties to stewardship of land and water resources. The comment by Havens, Chapman, and Irwin that campuses are "living laboratories for the latest thinking in green design and planning" rings true.[25] Most people look for green features in buildings such as solar panels or wind turbines. However, long-term sustainability can only occur at the planning level, beginning with a four-pronged approach:

1. Efficiency and Energy Conservation
2. Renewable Energy Supply Strategies
3. Resources Management
4. Alternative Transportation

Every major decision in the master plan must be informed by preserving water resources and lowering energy costs and emissions. Sasaki Associates' theory is spot on: "By tapping into the land grant legacy and other educational traditions, the nation might assume a leadership position in the environment as well."[26]

The University of Wyoming has developed a Long Range Development Plan. It has been a lengthy and tedious but necessary process if the university, a public land-grant research university, is to step forward and play a major role in sustainability policies for its region. Of course, economics will play a critical role in any long-term solution, and the university's plan recognizes the need to evaluate project costs and benefits not simply over a decade but over the forty-year period concluding in 2050. This plan is serving as UW's commitment to a long-range view of sustainability and as the starting point for next-level conversations on these issues at the university.

To be fair, even with the development of the plan, sustainability discussions occurring at the University of Wyoming are not where they should be. Universities and colleges that have taken initial steps such as purchasing hybrid cars for their fleet, increasing recycling efforts, and tackling carbon neutrality, for example, should be applauded as long as these same institutions treat sustainability not as a special or local interest, but as a topic that warrants global attention. To put it more simply, their campus conversations need to involve both demand-side and supply-side concerns so that their eventual recommendations are more than symbolic.

Conclusion: A Dose of Reality

The authors of the 2004 book *Limits to Growth: The 30-Year Update* modeled the consequences of a rapidly growing world population and finite resource supplies: "It is a sad fact that humanity has largely squandered the past 30 years in futile debates and well-intentioned, but halfhearted, responses to the global ecological challenge."[27] The road ahead will not be an easy one. While well positioned to frame substantive, results-oriented conversations, college and university leadership teams will face many challenges going forward, including at the University of Wyoming. For example, in April 2009, Wyoming's governor, Dave Freudenthal, took exception to comments made by Interior Secretary Ken Salazar that wind energy could replace coal-fired power. The initial comments made by Secretary Salazar at a public hearing put many contemporary opinions of renewable energy and sustainability in the forefront. Salazar was quoted by the Associated Press as stating, "The idea that wind energy has the potential to replace most of our coal-burning power *today* is a very real possibility. It is not technology that is pie-in-the-sky; it is *here and now*" (emphasis added).[28] Governor Freudenthal responded shortly thereafter that the prospect of wind power replacing coal in the nation's energy portfolio "ain't going to happen." Reflecting on the publicly reported exchange, Governor Freudenthal noted, "Public discourse on sustainable energy often includes a lot more rhetoric than reality. Conservation of energy and the mix of energy we, as a nation, choose to promote through government policy is critical to our collective futures. Equally important is that these discussions are grounded in reality. Government policy will not change the habits of consumers overnight, and making expensive energy alternatives more competitive through some type of government tax and subsidy arrangement is not without real transaction costs on our economy. Those transformations are not well served by statements of sustainability propaganda that fail to recognize the constraints of policy, technology, and our economy."[29]

Despite these challenges, higher education institutions can maintain a central role in achieving new levels of sustainability thinking nationally by (1) focusing on supply-side sustainability, (2) reexamining core values, and (3) broadening the present discussions occurring on campuses. The first step is to readjust the sustainability equation. As noted previously, campus leaders cannot address sustainability in a global way by focusing only on demand. The world's appetite for energy will continue to grow because of our drive for economic development, and the state of Wyoming, for its part, is spending tens of millions of dollars for the university to develop expertise in energy and resources. Over time, the American public will come to recognize that renewable energy resources are not the end-all fix and that the answer to this aspect of global sustainability is to use a combination of all the energy resources we currently have wisely and efficiently.

The second step is to reexamine our core values. The global trend has been to consume energy at an increasing rate for as long as we can. For some higher education community members, the response to this trend has been to market "sustainability" and "green," even if they divert critical time away from real research on a global problem. Real sustainability advances, however, will require more than changes in shopping habits. Eventually, the fad aspects of green will fade, and we will still need new and continuing sources of energy. Thus, to succeed over the long term, we will need to move from thoughtful considerations of to actual changes in behavior.

The last step is to broaden our discussions on campuses. At a time when the world is struggling to define a vision for the future of energy and the economy, sustainability conversations on campuses can be narrow and shallow. As resource-consumptive entities, colleges and universities have a responsibility to address the global energy crisis, in part, by teaching their students the real magnitude of the problem and not simply how to be more ecologically minded. Otherwise, many of our sustainability discussions are simply naïve, well-intentioned noise.

CHAPTER TEN

The Importance of Sustainability in the Community College Setting

Mary Spilde

Why the Moment Is Right for Community Colleges

To build a sustainable society that is focused on the triple bottom line—economically viable, socially just, and environmentally sound—community college presidents and leadership teams will need to take strong and immediate action. The "greening" of America is not a fad. It will continue to influence and affect our lives, which means that higher education needs to develop eco-conscious, highly skilled employees, students, and communities. This chapter explains the roles of community colleges and their leaders in these efforts, particularly focusing on the connections between community colleges and sustainability education.

Given all the challenges already facing community colleges in a climate of disinvestment and declining budgets, why is it important to engage in sustainability efforts? The answer is a simple one: mission. What sets community colleges apart from universities gives them a unique role in the higher education sustainability movement. Created to support and further social justice and service to community, the community college has sustainability at the core of its mission. The idea of the triple bottom line—economy, society, and environment—has direct application at community colleges, and as they continue to reconceptualize and reinvent themselves to meet community needs, sustainability is becoming the underpinning for a rising number of their mission expressions. Besides offering universal access, community colleges serve as local and regional economic drivers by providing a trained and educated workforce that fills the high-wage, high-demand jobs. This pool of relevant, practical knowledge forms part of the basis of what is needed to transition to a more sustainable society.

Six Lessons Learned at Lane Community College

1. Build on successes immediately.
2. Acknowledge the value of and energy in differences of opinion.
3. Resist the urge to take over as chief executive officer.
4. Unleash the power and passion of faculty, staff, and students without imposing as their president.
5. Start small, but start, and keep an eye on the whole system.
6. Tap into the collective intelligence of multiple campus groups.

Community colleges also serve thousands of students who transfer to universities. In his book *Ecological Literacy*, David Orr writes, "A genuine liberal arts education will foster a sense of connectedness, implicatedness, and ecological citizenship, and will provide the competence to act on such knowledge."[1] Helping citizens develop "the competence to act on such knowledge" is a specialty and strength of community colleges.[2] In addition, as the authors of *Cradle to Cradle* observe, "All sustainability is local."[3] Community colleges are typically deeply rooted and have extensive knowledge of and relationships with their local communities. "This connection to place makes community colleges especially well suited for education for new degree programs and extended community services in support of sustainability objectives" (see footnote 2).

Community colleges develop model training programs, including pre-apprenticeships, that ladder their curricula to take lower-skilled workers through a sequence of courses to provide a clear pathway to career-track jobs. As Feldbaum and States note in *Going Green: The Vital Role of Community Colleges in Building a Sustainable Future and Green Workforce*, community colleges are strategically positioned to work with employers to redefine skills and competencies needed by the green workforce and to create the framework for new and expanded green career pathways.[4] President Barack Obama's energy policies and focus on new models for a green economy identify the essential role of community colleges. Creating a sustainable economy is key to economic growth and prosperity, and central in this process is the role of public two-year colleges in training the new workforce and helping them to understand the growing impact of sustainability thinking.

The recently reauthorized Higher Education Opportunity Act includes the University Sustainability Program to strengthen existing campus sustainability initiatives. Two additional programs have been introduced in legislation. The Sustainable Energy Training Program for Community Colleges (2009) included in the U.S. Senate Energy and Natural Resources Committee would authorize $100 million a year for five years for workforce education and training grants to fund renewable energy and efficiency, green technology, and sustainable environmental practices. The Grants for Renewable Energy Education for the Nation Act (2009) provides support to develop career and technical education programs of study and facilities in the areas of renewable energy. The Campaign for Environmental Literacy is working to secure funding for these programs, dubbed "1 percent for education for a green economy."[5] This kind of investment would educate students about climate change and the green economy, including bolstering career pathways for youth. In all of the above, community colleges are perhaps the best-positioned higher education institutions to fulfill President Obama's agenda going forward.

The Middle Class Task Force chaired by Vice President Joe Biden proposed, in its October 2009 report, "Recovery through Retrofit," to expand green job opportunities by making homes more energy efficient. There currently are not enough skilled workers and green entrepreneurs to expand weatherization and efficiency retrofit programs on a national scale. This proposal will establish national workforce certification and training standards to qualify energy efficiency and retrofit workers and training providers. At Lane Community College, we work with community action programs on weatherization, but that is not enough. Students still need to be laddered into certificate and degree programs that lead to family-wage jobs. Community colleges, through their small business development centers, can help small businesses start up and expand—a natural for nurturing green entrepreneurs.

Three Emerging Trends

- Sustainability is moving beyond reductionist approaches to an understanding of the need for systems thinking and action.
- Collaboration and partnership is recognized as perhaps the best model to move sustainability forward—business, government, and local communities need to work with community colleges on long-term projects.
- Community colleges are key to leading and implementing the fundamental shifts required because of their reach and impact.

Three Best Practices

- Use the entire college as a learning laboratory for sustainability-in-action initiatives.
- Sustainability is a way of being, not an individual project—infuse it across operations, curriculum, and student services.
- Focus on the triple bottom line: the economy, the environment, and social justice.

Sustainability Thinking and Action at Lane Community College

Lane Community College saw the importance of raising the profile of sustainability-driven decision making on campus several years ago. Rooted in the south Willamette Valley in western Oregon, Lane is the second-largest community college in the state, serving more than 38,000 students. In fact, one out of every three adults in Lane County attends Lane annually. This gives the college a dramatic reach into and connection with its communities. In retrospect, Lane did not start with a policy, a plan, or a pronouncement regarding sustainability. Rather, the focus on sustainability emerged through collaboration. Faculty and staff who cared deeply about sustainability came together and started a dialogue. They did not wait for the State Board of Education or the college's administration to lead this effort. They began to imagine what a sustainable Lane could look like. Individuals started small projects. The important thing was that they *started*. By using a systems approach, we could view sustainability as a way of being, not as a project. In this process, the president's role was primarily to ensure a coherent institution-wide vision by planting seeds to foster creativity and mindfulness among community members.

As the initiatives emerged, Lane's vice president of operations was comfortable with ambiguity and saw her role as a living systems leader. She joined the conversation, provided support, and always kept an eye on the whole system. From there, working groups were convened to incorporate sustainability into both the curriculum and institutional operations. Cost-saving initiatives like recycling and energy conservation built momentum, and the savings were reinvested in other sustainability projects. Then performance benchmarks were established and sustainability was made part of several persons' jobs. Small and medium-sized successes were celebrated, promoting the good things that were happening.

Single Most Important Piece of Advice for a New Professional

- Do not let the complexities of sustainability thinking and action become an excuse to do nothing. Start small while keeping an eye on the whole system to assure that that whole is more than the sum of its parts.

This in turn created a reinforcing loop that led to even more initiatives and a gradual cultural change at Lane. Over time, sustainability thinking and action at Lane took on a life of its own. As described by Meg Wheatley in her book, *Leadership and the New Science,* new ideas and solutions were not planned, could begin anywhere, were unique, could not be transferred elsewhere, and were "leaderful" and irreversible.[6]

Campus Operations and Sustainability Goals

The college then began to implement a number of larger and more significant sustainability-driven projects in multiple operational areas. Even in terms of energy and water efficiency, we used a multifaceted approach. After a number of small projects, the college invested in an energy analyst position. This person is charged with auditing our energy usage, making recommendations for improving efficiency, working with local utilities on financing and incentives, and ultimately developing educational programs for our employees and students so that each of us is taking personal responsibility for saving energy. As a result, Lane has installed outside LED lighting across the campus, and when we renovate or construct new buildings, Lane adheres to LEED standards. The college is in the design phase for a new solar station that will not only provide electricity for the college and a charging station for electric vehicles but will serve as an opportunity for student learning. These examples demonstrate the integration of learning and operations in Lane's approach.

In 2006, Lane returned to using regular plates and silverware in its food service operation to limit the use of disposable products. The college composts its food waste, and its custodians beta-tested nontoxic cleaning and landscaping products. Set on 150 acres, Lane has returned much of the landscape to native plants to save on water and pesticide use, with the added benefit of bringing back monarch butterflies and other native species to the campus. The college hosts a student-initiated and -operated organic learning garden that supplies our food services

A Decision One Would Make Differently with Hindsight

- Move faster and make efforts even more visible to the campus community.

operations and our culinary program. Student farmers work with student chefs to grow local, organic food for use in the culinary and conference center. Finally, Lane is developing a Recycling Education Center that will be open to secondary students and businesses to learn about recycling and reuse and how practices can be implemented into daily personal and business routines.

Lane is fortunate to have a specialized support services unit at the college that provides employment training and education to adult students who experience developmental disabilities. The service operates as a cooperative venture between Lane Community College, the Lane County Office of Developmental Disabilities, and the State of Oregon Seniors and Persons with Disabilities Division. The service offers intensive individual and small group instruction that addresses social skill development, on-the-job employment skill training, work crew skills in socially integrated settings, supported work skills, and competitive employment placement. These students work throughout our campus in recycling, food services, and laundry and custodial services, earning a wage as well as receiving training that provides a level of independence for them. In addition to their valued work contributions to the institution, they also bring a diversity of viewpoints and experience and represent the college's social justice mission—a core element of sustainability action—at work.

A Sustainable Academic Plan for Community Colleges

New Career and Technical Training Programs

As the sustainability movement began to emerge, it appeared that there was a great deal of confusion about what a "green job" was and therefore confusion about how community colleges should respond. It is now clear that many of the green jobs span several economic sectors such as renewable energy, construction, manufacturing, transportation, and agriculture. While there will be new occupations such as solar and wind technologists, the majority of the jobs still will evolve from existing jobs. Thus, for community colleges this means developing new programs such as those outlined below to meet new industry standards, while also adapting existing programs and courses to integrate green skills. Exactly what new training programs a community college should develop depends on what is happening in its local community, yet it is wise to start developing a framework for green job training now in order to make informed strategic resource decisions about the curriculum at later points. Here are some programs that Lane College has considered:

—*Wave, biofuels, and geothermal technologies.* In these areas, either the technology is still undergoing development or there are few jobs in our region at present. We will keep these program areas on our horizon, but they do not need large scale program development at this point.

—*Wind and solar technologies.* Many companies already are investing in wind and solar, and these are two areas where community colleges are coming into their own, working with local economic development organizations to attract companies and then developing training programs to create a high-performing workforce.

—*Energy management.* Considering how much built environment we have in this country, it is apparent that this is where a vast number of jobs are now and will be in the immediate future. Lane has had an energy management program since the late 1980s that trains auditors to assess energy use and make recommendations and implementation plans for improvement.

—*Green job transformation.* There are few currently available jobs that environmental sustainability will not affect. Much of the necessary training for them will be accomplished through non-credit continuing education for the incumbent workforce and integrating sustainability curriculum into existing credit programs. The framework helped us focus on the areas that we felt would build on our existing strengths and develop new programs that would serve our region. In this regard, Lane added renewable energy technology, water conservation, and sustainability coordi-

A Sustainability Myth

- If I simply opt out of participating in sustainability actions on campus, they will go away like most other fads.

nator degree programs. The college's culinary program went through a reaccreditation by the American Culinary Federation in 2009, at which time it was informed that it was one of the top three sustainable culinary programs in the nation.

Curriculum Infusion

Community colleges like Lane need to educate students not just for a job but also for weathering a turbulent economy. It is essential for chief academic officers and the faculty, in particular, to look beyond job training and to integrate broader knowledge into the curriculum to empower students. Two-year degree programs already include general education, but if innovation is to be a foundation for our economic recovery, we must expose students to a curriculum that blends liberal arts and applied learning effectively. The college has an obligation to assure that every student, not just those in sustainability or energy programs, leaves with a clearer sense about how to create the triple bottom line of healthier communities, economies, and ecosystems.

While Lane offers Introduction to Sustainability, we recognize that not every student is able to enroll in the class and that we need to take a curriculum infusion approach. The faculty inventoried our existing courses and found that many disciplines had already integrated sustainability in their objectives. The college is now developing a process of adding an informational icon to our printed and online publications in the form of a "green stamp" to help students recognize that sustainability principles are infused into these classes. We are also working with faculty through a federal Environmental Protection Agency grant to educate faculty about sustainability thinking and principles and build their capacity to integrate it into their courses. The college's ultimate goal is to establish an eco-literacy requirement in all degree programs.

Learning Communities

Lane faculty members utilize a number of pedagogies affirmed by the Association of American Colleges and Universities as high-impact practices for successful student learning. For example, national research has shown that students who participate in learning communities succeed at higher rates than students who simply take stand-alone courses. While their format varies, a number of learning communities offered at Lane focus on sustainability, including "Ecotrails: Stewardship and the Sacred Landscape," which links Global Ecology with first-year writing, and "Reconnecting with Nature: Science, Spirituality, and Political Activism," which links courses in biology, political science, and religion. Service learning opportunities also provide students with work experience in community-based organizations that address larger community needs while promoting critical thinking, problem solving, and civic engagement. Experience in these three areas forms a core pedagogy of sustainability education.

Building a Sustainable Infrastructure

As many of these academic projects got underway, the college's leadership team turned its attention to developing the formal institutional infrastructure to support our sustainability programs and activities and to make them more visible. Lane signed the Talloires Declaration in 2004. This pledge made public our commitment to sustainability, and in 2007, Lane was one of the first signatories of the American College & University Presidents' Climate Commitment (ACUPCC). As the college began to actualize various institutional plans, the community also expressed interest in implementing a new Sustainability Core Value. The value, stated below, addresses the triple bottom line and is inclusive of staff and students:

1. Integrate practices that support and improve the health of systems that sustain life.
2. Provide an interdisciplinary learning environment that builds understanding of sustainable ecological, social, and economic systems, concern for environmental justice, and the competence to act on such knowledge.
3. Equip and encourage all students and staff to participate actively in building a socially diverse, just, and sustainable society, while cultivating connections to local, regional, and global communities.

Additionally, these Lane College Governance Council policies were adopted:

—Recycling policy

—Energy conservation policy

—Sustainable design and construction policy

—Transportation plan

—Design principles to guide master planning

Together with our academic and operational initiatives, policy and planning work has helped to embed sustainability into the core of college life. One key for Lane has been its previously mentioned systems approach that challenges Lane's leadership team to pay steady attention to all of the pieces that go on across the college and make sure they remain focused and connected. The culture of some colleges and universities requires beginning with institution-wide policy and planning processes rather than with specific projects. If that is the case, the Association for the Advancement of Sustainability in Higher Education (AASHE) has a wealth of resources available to member institutions that prefer this path forward.

Two additional features have been critical in the evolution of sustainability thinking at Lane. The first has been to encourage student leadership. Today's students have their own concerns about the environment, and they often expect that their college will not only understand the issues but will be proactive in responding to them. Students are no longer willing to settle simply for rhetoric: they expect to see tangible action. This provides avenues for student engagement and problem solving that make their degrees more relevant and enhance their learning outcomes. The second feature has been to use the college as a living laboratory. We have much at our disposal in the way of facilities, classrooms, and operational systems that can provide practical hands-on experience for students in all programs. For example, developing new sources of energy through the installation of solar panels has benefitted the college while simultaneously enhancing student learning. Using the college as a living laboratory also makes possible the development of a multifaceted sustainability program without the need for major allocations of funds in all areas at once.

Conclusion: Nine Action Steps for Institutional Leaders

It remains clear that the American public still has much to learn about sustainability issues and solutions and that higher education institutions need to assume a leadership role in addressing this concern through sustainability literacy courses, programs, and engagements. While there is no one path forward for a college or community college, there are multiple choices, depending on the institution's culture and context. Debra Rowe, President of the U.S. Partnership for Education for Sustainable Development, has been an outspoken advocate for the community college role in education for sustainability and has urged colleges to act now. The following nine concrete steps, created in consultation with Rowe, can mobilize an institution committed to sustainability goals and community action:

1. *Pursue strategic partnerships that include educational partners in K–12 and universities, economic development organizations,*

industry, labor, local government, and community-based organizations. These partnerships will allow community college leaders to serve as systems change agents and catalysts to educate the community about the triple bottom line and to facilitate necessary regulatory changes for a more sustainable future. Well-designed partnerships can also leverage and align public and private funding sources to enhance existing infrastructures, something that is essential in challenging budget times.

2. *Conduct a comprehensive greenhouse gas emissions inventory and develop a plan to reduce your institution's carbon footprint as soon as possible.* There are multiple models to assist with this available through the ACUPCC.

3. *Build on current strengths and capacities.* Ask the college's business development officer and senior faculty to study the local labor market and survey employers regarding the opportunities for existing and emerging green jobs and then work with department chairs to develop course and program offerings to meet these needs.

4. *Create technology-enhanced pedagogies so that students can learn at a distance, take general education courses at their local colleges, when feasible, and complete practicums in their local communities.*

5. *Embed sustainability in the campus culture and make the triple bottom line a core factor when making resource allocation, programming, and budget decisions.*

6. *Collaborate with national organizations like ACUPCC, Second Nature, the Association for the Advancement of Sustainability in Higher Education (AASHE), the Higher Education Associations Sustainability Consortium (HEASC), and the Disciplinary Associations Network for Sustainability (DANS).* For example, the AACC Sustainability Task Force, working with ecoAmerica, is developing a Sustainability Education and Economic Development (SEED) online resource center and online learning community to help community colleges access and share promising curriculum and practices, partnership models, and industry trend information.

7. *Act as a high credibility source of information for the public to increase their participation in energy audits, home weatherization projects, and greening their businesses.*

8. *Whatever its origin and level of success, tell your story.* Actively promote your institution's sustainability initiatives and green ca-

reer and technical programs. Besides earning positive media attention for the college, it will also continue to educate state policy makers, local business owners, and community decision makers.

9. *Finally, scale up.* Increasingly, in our competitive student marketplace, individual colleges are claiming to be *the* sustainable choice. Rather than waste resources by simply out-advertising each other, community colleges, in particular, need to leverage their progress and achievements and do what is best for their communities and the nation as a whole. We need to find new ways to scale up, and this can only be accomplished if we work in collaboration.

Over the past two decades, community colleges have become a key component of sustainability thought and action in higher education. In making their contributions to the promise of sustainable economic opportunity and prosperity for their communities, presidents' leadership teams, faculty, staff, and students are discovering that green decision making is sometimes simple, sometimes difficult, and always the right thing to do.

CHAPTER ELEVEN

Sustainability, Leadership, and the Role of the Chief Academic Officer

Geoffrey Chase, Peggy Barlett, and Rick Fairbanks

Framing the Campus Conversation

Growing concerns about the impact humans are having on our planet and the implications of those impacts on future generations have led many to argue that American higher education institutions have a key role to play in helping our society to meet the needs of the present without impeding the ability of future generations to achieve their needs as effectively.[1] Over the past twenty years, both scholars and activists have noted that there are critical domains through which higher education can contribute to a more sustainable world. Through the research they conduct, the engagement they experience with the broader community, and the operations they oversee, colleges and universities can serve as test sites for sustainable models and practices. Where colleges and universities may have the largest impact, however, is with the students they educate. As David Orr has argued, the real challenge we face in embracing a more sustainable future rests with our ability to educate students differently.[2]

Today there are more than eighteen million students in colleges and universities in the United States alone, and if they graduate with the needed skills to help societies develop more sustainably, higher education will have fulfilled one of its most important roles in leading us in a new direction.[3] More specifically, graduates will need the skills, knowledge, and habits of mind that prepare them to meet the challenges presented by climate change, loss of biodiversity, a world population of nine billion in 2050, decreasing water resources, global health challenges, and extreme poverty. Academic deans and vice presidents for academic affairs will have key roles to play in effecting the kind of changes within higher education that are needed. Next to presidents and chancellors, chief academic officers (CAOs) hold the most important leadership positions within a college or university. In addition to overseeing all aspects of the academic program for the institutions they lead, they are "first among equals," working side by side on a daily basis with vice presidents for advancement, student affairs, business affairs, and technology.

CAOs need to exercise their leadership authority related to sustainability in ways that will maintain and extend the relationships with those who report to them, those who work as colleagues, and those to whom they report. While the leadership position CAOs hold provides them with opportunities to promote sustainability and to help their institutions shift to support the activities already noted, these same CAOs often meet significant challenges. Chief among these is that colleges and universities are "loosely coupled" organizations in which the different parts of the organization retain their own identity and separateness.[4] Thus, a CAO must enlist others in moving toward sustainability collectively. This is true not only for issues of sustainability but also for those of academic leadership in general, including assessment, increased faculty research, more rigorous academic programs, a greater emphasis on high-impact educational practices, and internationalization. While each of these is a significant undertaking, for CAOs, sustainability presents a special set of challenges.

Foremost among these is that sustainability reflects a paradigm that, as one CAO has noted, is challenging because if "sustainability is everything, it is nothing."[5] As John Tallmadge has argued, "Sustainability is not a problem, a condition, or a program; it's a way of life." Thus, CAOs who wish to promote sustainability may wonder where to begin.[6] Put simply, it is hard to measure. In those areas of campus life where sustainability practices may be more easily measured, such as in operations, the CAO will be wise to assess collaboratively with the vice president for business affairs or operations. Remembering that sustainability manifests itself at the intersection of the domains noted above—research, engagement, operations, and curriculum—CAOs committed to promoting it will need to remain engaged in all aspects of the campus and should plan to work closely with colleagues who hold responsibility for the nonacademic spheres of the institution.

A second challenge lies in the fact that the curriculum is driven by faculty knowledge and expertise. Faculty committees oversee curriculum changes, decide whether particular courses are appropriate for general education programs, and approve new degrees, majors, minors, and certificates. Although the chief academic officer has responsibility for the academic program at his or her institution, the faculty develops and has primary responsibility for the curriculum. A third challenge is that although chief academic officers have strong leadership positions, they may also feel somewhat constrained as they negotiate changes within the existing culture of their institutions, particularly in periods of scarce resources. Recognizing that institutions vary dramatically in terms of culture and mission, this chapter will provide specific examples from a range of colleges and universities to illustrate the roles CAOs can play in helping their institutions become more sustainable.

Over the past two decades, many faculty members, academic administrators, and other educational leaders have worked collaboratively to defend what it means to teach sustainability. One of the most useful efforts to do so has emerged from the Curriculum for the Bioregion initiative of the Washington Center. Recognizing that sustainability involves "learning to make decisions that provide for the needs of the world's current population without damaging the ability of future generations to provide for themselves," the center has provided several key sustainability concepts. These include (1) interconnectedness, interdependence, and systems understanding; (2) equity and justice; and (3) global to local perspectives or biospheric to bioregional perspectives as well as a list of the sustainability "habits of mind" that students need to develop.[7] In a similar vein, AASHE has adopted a goal suggesting that students be able to synthesize an understanding of social, environmental, and economic forces and be able to apply that understanding to real world problems.[8]

Sustainability is inherently interdisciplinary, and often the structures of our colleges and universities do not support forms of interdisciplinary teaching across departments or even within particular courses or areas. There is also no consensus as to whether sustainability should exist as a stand-alone program and within its own set of degrees or if it should, as Tony Cortese has argued, be integral to every discipline.[9] If the aim is the former—sustainability programs—then CAOs must ask what is the necessary academic preparation for faculty teaching in such a program? As important, however, are the questions one begins to ask when considering that sustainability increasingly belongs in *every* aspect of the curriculum. What does it mean to integrate sustainability into disciplines such as English, accounting, political science, Chicano and Chicana studies, graphic design, marketing, and so on? Just as there is no single definition through which we can know exactly what constitutes sustainability, there is presently no national, much less international, agreement about what constitutes sustainability education. Moreover, the curricula offered in most universities and colleges today do not typically have one goal or one series of intentions in mind. Integrating sustainability into curricula for workforce training, for example, will be different than integrating sustainability into a liberal arts curriculum in the general education program. The skills and abilities related to sustainability that particular majors need will vary from discipline to discipline.

The CAO as Sustainability Advocate and Advisor: Three Critical Steps

In spite of these challenges and this complexity, CAOs play key roles in helping their institutions adopt models of education that shape education for sustainability. Furthermore, while institutions do differ in regard to mission, scale, and funding, three overarching perspectives cut across the curricula of almost all colleges and universities: (1) building on current commitments, (2) focusing on place, and (3) supporting faculty and curriculum development. This chapter provides a review of these perspectives while illustrating how they manifest themselves in different settings.

Build on Current Commitments

For a new dean or provost, sustainability, as important as it is, can seem like just one more helping on a plate that is already very full. Budgets, enrollments, promotion and tenure decisions, new faculty hires, institutional accreditation, assessment, and working with colleagues in other divisions across the college or university already constitute a full-time set of tasks. Similarly, faculty—especially young, tenure-track teachers—also feel overworked as they are asked to take on "one more thing." One way for a CAO to address this workload issue and any associated resistance to the idea of sustainability itself is to reframe the discussion so that it focuses on current commitments.

Faculty members at small colleges, for example, have a deep sense of mission and a commitment to both the explicitly stated and implicitly understood role that the institution plays in the education of its students. The sense of institutional identity on such campuses is exceptionally strong and shapes how the faculty perceives the need for and the possibility of change. Thus, whether CAOs emerge from the faculty of the college or arrive from a position at another college—both of which present differing sets of opportunities and challenges—those at small colleges must acknowledge, learn to negotiate, and work within the community as they find it. This presents challenges because change may represent a challenge to the core mission of the college. However, at institutions like Northland College, College of the Atlantic, Green

Mountain College, and Unity College, the core mission became an opportunity because liberal education done traditionally is perhaps education for sustainability.

How Northland's new liberal education program, "Natural Connections," came to be is instructive for a number of reasons. First, while the program contains an explicit sustainability outcome, it was not constructed *as* a sustainability curriculum. While the notion of sustainability is widely employed at Northland, the concept played virtually no role in the construction of the "Natural Connections" program or of the (Lake) Superior Connections alternative. If those programs are sustainability curricula in virtue of satisfying the description given above, then it follows, interestingly and perhaps surprisingly, that sustainability curricula are not necessarily *about* sustainability.

In some respects the integrated, holistic, pedagogically engaged aspects of sustainability curricula have always been part of the classic "education for citizenship" account of liberal education. In fact, the only aspect of sustainability curricula missing from that account is the requirement of ecological knowledge.

What makes this connection to the liberal arts consistent with sustainability is a series of core courses that, while addressing the essential learning outcomes as articulated, for example, by the Association of American Colleges and Universities (AAC&U), is attention to environmental issues that ensure that education for sustainability be environmental or ecological. Core courses in Unity College's "Environmental Stewardship Core Curriculum" include "The Environmental Citizen," "Environmental Sustainability," and "Environmental Challenge." Green Mountain College's "Environmental Liberal Arts Core Sequence" includes "Images of Nature," a course that "explores some of the ways human societies make sense of the natural world"; "Dimensions of Nature," featuring scientific approaches to nature; and "A Delicate Balance," which raises explicitly the question of citizenship.

At the College of the Atlantic, all first-year students take the Human Ecology Core Course. At Northland, the general education curriculum, "Natural Connections," is almost entirely composed of integrated sets of courses, one four-course block and two two-course blocks. The four-course block, "Foundations in Nature," taken by all first-year students, requires emphasis on "the use of scientific methodologies to investigate natural phenomena" and "the examination of relationships between humans and the natural world." These approaches to general education reflect an emphasis on integration and the centrality of environmental knowledge and illustrates that as we seek to educate environmentally responsible world citizens, sustainability is consistent with general education.[10]

Northland College: How New Curricula for Sustainability Happened

RICK FAIRBANKS

Since there is always an expectation that the new dean or president will attempt to stamp a place in his or her own image, the Northland case considers the impact of outsiders who are able to call institutions back to their mission and history precisely because they are outsiders. If the Northland case is instructive, it shows that the role of a new chief academic officer or president can be akin to the role of the convert who, by virtue of newness to a tradition such as that of sustainability thinking, sees it with eyes that discern possibilities, virtues, and assets that the habituated may overlook. This is not an argument that home-grown provosts and presidents cannot lead revolutionary change in colleges and universities, but rather it illustrates that newly hired leaders need to think carefully about how to leverage their outsider status to help their new institution see with freshness and clarity its own traditions.

A crucial point in the change process of Northland occurred when the new vice president for academic affairs challenged the faculty by offering to begin working on his own approach to implementing sustainability in the college's curriculum. The faculty immediately came to him with the warning that acting unilaterally would make the CAO's leadership, or lack of it, the issue and prevent real discussion of sustainability going forward. The faculty's response was crucial because from that moment forward, it was acknowledged that Northland's professors now owned the changes to their curriculum.

The following list summarizes the background factors challenging the college's academic leaders as they developed a stronger core of sustainability thought and action within the community:

- Northland did not have curricula that clearly explained or reflected its mission. Neither prospective students nor their parents could discern what "environmental liberal arts" meant by looking at the original mix of academic programs.
- Northland's back was against the proverbial wall in terms of declining enrollments, and since faculty said that they were the heart of the institution, it was incumbent upon them to lead the institution forward through these curriculum revisions.
- Northland did not have curricula that took advantage of its distinctive market niche and reputation. It was trying to compete with better-resourced, better-known, "generic" liberal arts colleges on their terms.
- The proliferation of the college's very small, one- or two-person departments, raised continuing questions about depth and quality.

(*cont'd*)

Northland College: How New Curricula for Sustainability Happened

- Building interdisciplinarity into majors was a necessary innovation in order to avoid the instability of entire programs of study being delivered by one or two people.
- The vice president for academic affairs made it known that at the end of the process, status quo would not be acceptable to the president and trustees.
- Over time, as the faculty made steady progress in focusing the curriculum on sustainability themes and thinking, the chief academic officer became less of a "challenging outsider" and more of a resource-finding—and allocating—supporter and colleague.

If the curricular changes prove to be the right ones for Northland, they can be said to be a victory for shared governance. The vice president for academic affairs was able to play the roles described above because a group of dedicated faculty leaders successfully managed much behind-the-scenes coalition building and, more importantly, supplied the ideas to refocus and refresh a body of courses and degree programs that had lost their distinctiveness among thousands of student choices. This initiative was followed by a faculty Transitional Task Force that recommended a strategic hiring plan to staff the new curriculum and that helped the college to expand and strengthen its expertise in fields such as sustainable community development, ecopsychology, and Native American environmental perspectives as additional benefits.

Many institutions also recognize the role that high-impact educational practices play in furthering student development, and research indicates that these practices are especially beneficial for historically underserved students.[11] All of these practices—first-year seminars, common intellectual experiences, learning communities, writing-intensive courses, collaborative assignments and projects, undergraduate research, diversity/global learning, service learning and community-based learning, internships, and capstone courses and projects—are consistent with sustainability education and thus do not represent a departure from current commitments but a joining of those commitments with sustainability. At San Diego State University, for example, students are engaged in seminars through which they can explore interdisciplinary topics related to sustainability and engage in applying what they are learning to real-world challenges. These opportunities have made it possible for students studying the arts community in Imperial Beach, California, a diverse city bordering Mexico, to spend some of their weekend time with local artists to explore how a commitment to the arts contributes to the sustainability of community. Other students have travelled regularly to the Santa Margherita Ecological Preserve in northern San Diego County to help establish a low-water-use organic garden that will supply food to the dining halls on campus. Each of these activities is tied to academic course work and thus helps to extend beyond the walls of the classroom what students learn. Thus, education for sustainability is already consistent with best practices for undergraduate education and the commitment to preparing students to address challenges they will face.

Focus on Place

As faculties at colleges and universities have taken up a commitment to sustainability education over the past twenty years, a focus on place has emerged as an enduring theme. Just as it can be useful for a CAO to support current commitments and their linkages to sustainability, it has also been very useful for CAOs to support a renewed and consistent focus on place, region, or area. As John Tallmadge has noted, "Sustainability must always manifest itself in some place with some people; it always has a local, personal flavor."[12] Knowing the place where we are, recognizing its distinctiveness, and suggesting that both faculty and students use it as an opportunity for learning about sustainability has provided a powerful set of connections on many campuses. Connections to *place* bring the invisible surroundings of the campus into focus and help participants see how economy, environment, and social justice are inextricably linked.

One strategy for a provost or vice president for academic affairs is to ask her or his institution to revisit, or focus on, the place, area, or region it occupies. Many colleges have a clear identity and mission tied to the area in which they are situated. Northland has offered integrated place-based curricula to Outdoor Education majors for more than thirty years. "Fall Block" combines natural history, adventure education, environmental education, and expedition planning in a semester-long program at the Audubon Center of the North Woods. The college launched another place-based curriculum two years ago in the "Superior Connections" program, a two-year, nine-course sequence focused on the Lake Superior watershed that fulfills all of students' general education requirements. In 2010–2011, Northland launched another integrated alternative to its general education curriculum focused on sustainable agriculture, with multiple hands-on experiences with local agricultural initiatives called "Growing Connections." Green Mountain's block courses are multidisciplinary, multiple-credit, semester-long approaches to topics such as "The Vermont Wilderness," "The Hudson River," and "The Champlain Basin." The Eco League is a consortium that provides students at five colleges—Alaska Pacific, College of the Atlantic, Green Mountain College, Northland College, and Prescott College—the opportunity to take part in exchange programs focusing on environmental responsibility on those campuses. Thus, place, wherever students may be studying, becomes an opportunity to make significant connections around sustainability issues.

A focus on place has also emerged at larger universities. At Michigan State, for example, faculty, realizing that sustainability awareness needed to be grounded in place, developed a comprehensive lecture course, "Our Place on Earth: Experiencing and Expressing Our Relationship to the Natural Environment." Student writing exercises explored readings, lectures from visiting writers and faculty scholars, and the campus (the cyclotron, laundry, bakery, research farms, and power plant) and environs (wood lots, gardens, and the Red Cedar River). "Students saw connections between the outer space of astronomy and the inner space of nuclear physics. They questioned the ethics of high-tech agriculture and the absence of local food in the dining halls. They found parallels between the loss of natural areas on campus and the loss of local businesses on Main Street."[13]

Many faculty members find themselves teaching where they did not grow up or attend school. The focus on place often has a powerful impact on faculty and can serve as a way for provosts and academic vice presidents to provide support for faculty as they connect their disciplinary background with the campus and region and with their teaching. At one sustainability workshop, a presenter at the University of Scranton in Pennsylvania used a walk along the nearby river to highlight the history of coal, immigration, and labor struggle in the region. After he pointed out a visible coal seam on the riverbank, he linked it to continuing environmental impacts from mining upstream and then pointed to a nearby street intersection where, decades before, protesting miners were shot by deputized town leaders. Most faculty attenders at the workshop did not know that history or how the mineral wealth of the area was bound up with continuing environmental issues. The historical example of struggle for social justice connected with the current regional economy. The long-term environmental consequences of resource extraction stimulated thoughts about teaching those same kinds of interconnections in the context of current Pennsylvania culture.

A Different Shade of Green: A Junior Faculty Member Listens to Administrative Sustainability Speak

TIMOTHY FARNHAM

As a junior faculty member in the Environmental Studies Department at the University of Nevada–Las Vegas in 2008, I was pleased when our senior administrators caught sustainability fever. The president mentioned it in some form in every major address, colleagues were getting grants for projects advancing sustainable research, students were calling for more courses exploring sustainability, and staff members were asking the university to become more sustainable in its operations. Indeed, sustainability was probably the second hottest topic in higher education after shrinking budgets and endowments, and faculty members—especially young assistant professors—felt the pressure to hop on this green bandwagon. Internally, the central question for many of my colleagues was a basic one: What does the administration mean by "sustainability"?

From my own personal experience, I remember thinking, "Well, our department has been studying and advocating for sustainability since the beginning." I found it amusing that the administration had just discovered this new topic that many of my colleagues had known about for years, but as I sat in on the second annual UNLV Urban Sustainability Conference and listened to a university official speak about the university's commitment to sustainability, it became apparent that he was using the term much more broadly than I ever had. The projects he listed were all noteworthy: removing turf to save water, raising recycling rates on campus, assessing chemicals used in laboratories, and increasing the mix of renewable energy in the university's power portfolio. Yet this talk of sustainability gradually but clearly took a different turn and became one of "economic sustainability" of the community and the role that the university would be playing in enhancing the surrounding business environment. I listened with some uneasiness. We had moved from environmental projects to development projects that contained many fewer plans for being environmentally friendly. I became confused about what the administration's position was on sustainability. If a sustainability project was not going to save the university money, would it still be considered next year? Were development projects that helped the local economy to be put in the same sustainability category as major environmental improvements?

For this reason, as a junior faculty member considering what I had to do to earn tenure, I initially became cynical about the general university definition of sustainability. As a dedicated environmentalist, I still valued what I saw as the original concept of sustainability: the judicious use of resources that did not impair the quality of life for present and future generations. Yet now that the president and trustees had taken hold of the term and were using it to advance

(*cont'd*)

multiple agendas, this gave me little direction. How was I supposed to know how to support the university's sustainability mission if I could not define it clearly enough for my department and myself? What I needed was a more detailed statement of UNLV's intentions.

Then I read the following in the conference program: "UNLV is now playing a major role in establishing a dialogue with the Las Vegas community on all of the interrelated dimensions of sustainability, including environmental, economic, and social/cultural sustainability."[a] I realized the advantage of this immediately. In essence, the UNLV board, president, and senior academic officers saw the sustainability wave as an opportunity to join several key messages more efficiently:

—Make the university greener and support greater numbers of environmentally friendly policies and projects designed to alleviate global and national crises

—Sustain and then grow the university's financial resources, in part to be able to implement more sustainability-driven projects and partnerships

—Commit to maintaining sustained economic health of the Las Vegas community surrounding the university and also to achieving expanded social and cultural diversity goals which have continued to form a key element of UNLV's mission in Las Vegas and Nevada

As skeptical as I was when I heard how the administration had decided to stretch the concept of sustainability, I had to admit that it was an effective vessel of communication for the institution's long-range goals and agenda. In addition, when one looks at the diversity of related projects on the UNLV campus, it is clear that faculty and staff members are not confused by these multiple presentations of sustainability. Rather, it appears that they have taken the administration's lead in considering sustainability as a growing cultural value with multiple dimensions and as a kind of umbrella for numerous useful university initiatives.

Timothy Farnham is the Leslie and Sarah Miller Director of the Center for the Environment at Mount Holyoke College. Prior to that, Farnham served as Associate Professor and Director of the Undergraduate Program in Environmental Studies at the University of Nevada, Las Vegas.

[a]Conference program, Urban Sustainability Conference, "Education for a Global Future: 21st Century Challenges in Sustainability & Climate Change Education," University of Nevada, Las Vegas, March 6, 2009.

Similarly, faculty members at San Diego State University are intrigued to learn how several murals painted on campus walls as part of the Works Progress Administration, and hidden for decades, link the history of the campus with the tuna industry in San Diego, immigration patterns in southern California, and the New Deal. Place plays an important role through experiential learning, a powerful component of sustainability education, for both faculty and students. Time in natural spaces opens our eyes to the living systems around us and the human impacts on them—invasive species, polluted water, eroded stream banks—and is an important step in galvanizing new connections in unusual classes.[14] Helping faculty rediscover the uses of outdoor experiences can empower them to carry out their own field trips, a big leap for some. One faculty member said, "Campus life is almost entirely indoors, office and computer oriented—this project forces me to think about where I am and how my actions are part of a natural and human ecology."

Support Faculty and Curriculum Development

As we seek to produce students able to construct a more sustainable future, it helps to have faculty who have experienced their own intellectual growth and personal creativity around sustainability. An expectation that faculty can be encouraged to integrate environmental and sustainability issues in the curriculum entirely through self-education is probably unrealistic.[15] Thus, for provosts to accelerate faculty commitment to renewal and change, some form of curriculum development program will be necessary.[16]

Inspiring faculty creativity is the key to integrating sustainability into the curriculum, and providing adequate professional development resources for faculty to succeed in this task is finally the responsibility of the campus CAO. Creativity requires a trusting, safe space to assess current practices, admit a willingness to change, and consider alternatives. The Piedmont Project at Emory University, based on the Ponderosa Project at Northern Arizona University and the Tufts Environmental Literacy Institute before that, has developed one model of a workshop and faculty development format that has been successful in creating communities of trust and has been adopted by a number of other schools around the country.[17] These workshops, as well as the workshops offered by AASHE on Leadership for Curriculum Change, are based on the recognition that faculty are responsible for the curriculum.[18] Each professor builds on his or her own expertise to offer students a unique experience in each course. The challenge for sustainability education, then, is to stimulate the desire for change within the broadest possible group of faculty.

Sustainability challenges demand broad linkage among issues, which can create some discomfort for faculty members who want to have all the answers before they go into the classroom. Education for sustainability often involves extending one's interests into new areas, which can be fun but also a bit disconcerting.[19] A willingness to step outside the "expert" role, into a stance of co-learner with students, is a disposition that cannot be demanded but must be nurtured and rewarded. Said one faculty member, "One of the best benefits I've seen in Piedmont Project is that it provides a forum for people to talk, learn, without needing to be 'the expert.' It's a place to actually be safely curious." Paradoxically, when teachers give up a slice of the expert role, it can support leaps of learning, for both students and faculty. The trust built during a good faculty development program is critical to the shift in paradigm and approach that will allow the most creative pedagogies of the future.

Building intellectual excitement toward sustainability education begins with mixing faculty from diverse fields in a context that downplays rank, power, and prestige. One important part of creating a safe space is signaling that all participants are equally worthy and valued. It is useful to accept participants at all career stages and to encourage a range of curricular projects, those centrally and obviously connected to sustainability (e.g., a philosophy course on Thoreau or a science lab on green chemistry) as well as those further afield (a math course that redesigns ecological footprint calculators or a science course that includes political and economic dimensions). Presentations make clear that scientific or technological contributions join many others in value; no one area of the curriculum "owns" sustainability.

As advocates for effective teaching of sustainability, deans and vice presidents for instruction need to call for teaching and learning activities that transcend the disciplinary. The attempt to transcend disciplines is particularly evident in some of the institutions previously mentioned in this chapter. The College of the Atlantic, for example, offers just one major, Human Ecology, and has no departments. The College of the Atlantic's faculty is organized into "resource areas" rather than departments, each composed of faculty from multiple disciplines. Unity College's faculty is organized into centers that hold faculty from multiple disciplines, as is true of Northland's new departments of Environmental Sciences, Nature and Culture, and Social Responsibility. Grouping faculty from multiple disciplines does not guarantee interdisciplinary work, but it removes one barrier to conversations that lay the groundwork for such work. Cross- and trans-disciplinary majors in sustainability fields can be found at Northland, Unity, and Green Mountain: Green Mountain's Environmental Studies, Environmental Management, and Natural

Resources Management majors as well as their graduate degrees in Environmental Studies and their certificate in Sustainable Business; Northland's Humanity and Nature Studies, Sustainable Community Development, Environmental Geosciences, and Natural Resources majors; Unity's Environmental Humanities, Environmental Analysis, and Agriculture, Food and Sustainability. Although the College of the Atlantic offers only one major, it does feature "focus areas" in Sustainable Business and Sustainable Food Systems. Returning to an earlier point, it is probably best to understand this array of interdisciplinary programs as experiments in education for sustainability.

Other strategies to build curricular engagement include *faculty learning communities,* reading groups, or seminars. A good approach pairs a semester of such activities with a team-taught course the following semester, thereby fusing content learning and curricular development. Workshops to help faculty prepare for new sustainability general education requirements, majors, or minors can be carried out in a way that builds a trusted network of supportive colleagues, not just independent course change. Similarly, *field trips* are a useful strategy to explore local issues and can be very effective in providing opportunities for faculty to work across disciplinary lines with each other to consider the relationship of sustainability issues to their teaching. The camaraderie of travel, the hands-on experience of getting to know issues in the nearby area, and the sense of "time out" from the usual all combine to stimulate new ideas. In one half-day field trip, an Emory group visited a low-income neighborhood in Atlanta devastated by frequent sewer overflow contamination and then toured the municipal water treatment plant. The social, economic, and environmental aspects of the water system of which we are a part were a powerful learning opportunity. Several faculty members were inspired to teach an interdisciplinary course on water, exploring geological, chemical, literary, philosophical, biological, musical, public health, and international aspects of water. The courses were extraordinarily successful in enabling students to think in complex ways about some critical issues in their region.[20]

At the same time, it is also important for provosts and deans to reassure their faculty that integrating sustainability does not necessarily mean large-scale changes to a course. Nor does it mean that every faculty member must develop whole new areas of expertise. Instead, if many faculty members make changes, even relatively modest ones, these changes, taken together, make a significant difference in what students learn. It is for the chief academic officer to learn when to apply pressure and when to trust the process. Sometimes the most effective change is simply in refocusing an existing assignment. A nursing course on lactation added an exercise in which students develop a patient assessment of the pros and cons of breastfeeding for U.S. and international contexts. The workup took into account cultural norms, the mother's obligations to work outside the home, and childcare arrangements as well as the availability of potable water, energy and other resources used in manufacturing and transporting formula, bottle sterilization, and waste disposal. The exercise highlighted the often-ignored costs of bottle-feeding, while taking care not to prescribe what should be the mother's decision.

Team teaching and *guest lectures* also emerge as ways to integrate sustainability into a broad range of courses. Relationships with nature in early Western thought—and their implications for how we live today—led to collaboration between a classics professor and a philosopher, in a course on Concepts of Nature. A religion professor teaching about contemplative practice welcomed an environmental studies colleague to give students more scientific knowledge before a meditation field trip in the woods. Some faculty members begin by assuming that sustainability issues can be inserted into an existing course and then come to realize that the whole course paradigm has to shift. Often this way of bringing sustainability into the syllabus is the most powerful of all, because it illustrates for both the professor and the provost that sustainability is, in fact, a paradigm through which they can examine most syllabi in most disciplines. A number of business schools nationally have now framed their curricula with sustainability's "triple bottom line," thereby broadening the definition of business success.[21] An anthropology course also used the triple bottom line as an organizing theme for a course on wildlife conservation. In another course on medieval art history, a faculty member began asking her students to look at the art they were studying while asking, "What is the connection between art, faith, and the environment?"[22]

Conclusion: The CAO as Sustainability Activist and Resource-Builder

Although many of the examples discussed in this chapter have highlighted faculty innovations, they have almost always emerged with and because of support from the CAO and the academic administrative team. CAOs have a critical role to play on their campuses as initiators, supporters, and intellectual leaders. Increasingly, students, employers, and the broader public are seeking programs that prepare future generations with the skills, abilities, and knowledge they will need to build a more sustainable society. A provost or dean's commitment to sustainability supports the broader social contract that institutions have as they help build new foundations for the future. As general and philosophical as these commitments may be, there are related specific and concrete actions that CAOs need to take to provide support and leadership.

First, and most importantly, provosts will need to make a difference through tangible support for faculty development programs that provide avenues through which faculty can learn, interact, and develop their own approaches to sustainability. Faculty development programs create communities of trust. These communities then lead to rich opportunities for dialogue across disciplines and implicitly support faculty members as they learn from each other. Finally, these faculty development programs generate strategies that can be shared across campus to help faculty colleagues consider changes to their own courses while simultaneously considering how their teaching fits into the overall curriculum.

Second, regularly raise sustainability issues with colleagues in student affairs, advancement, athletics, and business affairs. The efforts to help institutions support sustainability are strengthened when they involve and coordinate the multiple operational arenas of the campus. As these institutional offices integrate sustainability into business practices, student affairs programming, and sustainable athletic events, new opportunities will emerge for student learning.

Third, take time, as the CAO, to learn more about sustainability not only at the campus level but also within your own original discipline. Since many CAOs today did not receive their professional training when sustainability was a key priority, it can be both useful and instructive to consider, as faculty colleagues across campus are doing, how one might teach something about sustainability in one's own field of research.

Finally, present sustainability as a series of high-profile opportunities for celebration and for extending the current commitments of your institution. Sustainability represents a new paradigm, but it is a paradigm that will be successful only as it helps us reconnect to the places we inhabit, to all the communities of which we are a part, and to the core commitments and values reflected on our campus and in higher education around the world.

III

Fresh Agendas for Campus Operations

CHAPTER TWELVE

Greening the Endowment

Mary Jo Maydew

As the concept of environmental sustainability becomes more deeply understood and widely accepted as a key institutional goal by colleges and universities, it is being applied to areas beyond those initially associated with becoming "greener." One of those areas is the management of the institution's invested funds. Applying the principles of environmental stewardship to investing may seem less obvious than green building design or energy conservation, but green investing is one more way an increasing number of higher education institutions are achieving their visions of environmental sustainability.

Green investing is a growing component of socially responsible investing, which has a long and deeply rooted history among many college and university presidents, financial vice presidents, and trustees, in particular. This model of investing is defined broadly as the use of a process for selecting investments that considers social, environmental, and internal corporate issues, including governance and executive compensation, as part of the decision-making process. The roots of shareholder activism go back to the 1930s. The earliest issues were financial ones—expansion of financial disclosure, increased dividends, and management accessibility. A second wave of activism arose from the social, environmental, and political agendas of the 1960s and 1970s, and included issues such as civil rights and nondiscrimination, anti-war activities, pollution, and consumer safety. By far the most enduring issue, and the one with the broadest engagement, was the growing concern about companies, particularly U.S. companies, doing business in apartheid South Africa, that began in the 1970s and ended only with the dismantling of apartheid in the mid-1990s. The creation of the *Statement of Principles of US Firms with Affiliates in the Republic of South Africa,* written by the Reverend Leon

Sullivan in 1977 and popularly known as the *Sullivan Principles*, provided a carefully considered template for determining companies' performance in South Africa and was used by organizations, including many colleges and universities, as a guideline in considering whether to make an investment, or continue to invest, in such companies.

While investing in companies doing business in South Africa has received the most attention and perhaps had the most influence on companies, socially responsible investing addresses a much broader range of other matters, from worker exploitation (e.g., anti-sweatshop measures and intimidation of pro-union activities), to health concerns like marketing highly sugared cereals to children and the production of tobacco products, to corporate governance issues such as golden parachute provisions and the percentage of independent directors. The focus on environmental and sustainability concerns goes back almost to the beginning of the social responsibility movement and was accelerated by the Exxon Valdez oil spill in 1989. Sustainability thinking with respect to investing runs the gamut from concerns about environmental stewardship being integrated into corporate missions to supporting the development of specific green products and methods.

Proponents of socially responsible investing make two arguments for its value. First, for those institutions that already practice some form of socially responsible investing, it is viewed as the right thing to do—a matter of social justice and making the world a better place for future generations. Second, colleges that believe in some form of screening will argue that companies that continue to produce damaging or potentially dangerous products or that are not actively working on issues of environmental sustainability will eventually become poor investments. Although typically not among the voiced arguments for socially responsible investing, a third reality for some colleges and universities is political: student and faculty pressures to take a stand for or against a particular corporation's behavior.

As more higher education institutions openly embrace environmental sustainability, the primary rationale for enlightened investing will likely move gradually from the first to the second justification, that is, from a social justice and philosophical lens to an economic one, as companies and funds that are not environmentally responsible will become decidedly less-attractive investment opportunities.

Considering Green Investing for Your Institution

How green investing is defined varies widely, and each university and college needs to consider what it will mean to that institution. Par-

Three Best Practices

- Create a community process that is congruent with the institution's environmental goals and priorities.
- Start gradually and educate the community about green investing before launching major initiatives.
- Use environmental impact as part of the process of researching and selecting investments.

ticularly for institutions that are just beginning to consider whether and how to apply an environmental screen to investing, several baseline questions need to be asked of its mission: Does the college want to encourage and support companies developing renewable energy technology and producing green power? Does the board's investment committee want to dissociate from companies it determines to be acting in ways that systematically damage our environment? Are members of the leadership team concerned about the proliferation of genetically engineered foods or the use of certain pesticides by food production companies? Is the sustainability coordinator advocating support for the local production of goods and services? Is the chief financial officer concerned about broader access to capital through micro-lending or community-based development activities? These are only a few examples of numerous issues that institutions will need to consider.

In addition to reaching some agreement on how broad or how narrow the environmental lens will be, it is also important to think about the process that the university or college will use, how widely it will consult with its campus community, and how actively it will remain involved in assessing outcomes. While the approach may evolve over time, a shared understanding of process is key to engaging the community productively in socially responsible investing. Some institutions manage the process centrally by developing a policy statement that includes socially responsible investing and invites community participation in its drafting while leaving implementation to the investment committee of the board of trustees and the chief investment officer. Other universities retain a representative constituent committee that provides advice on various topics. Harvard University has an Advisory Committee on Shareholders Responsibility composed of faculty, students, and alumni that focuses entirely on making recommendations on the voting of shareholder resolutions.[1]

Carleton College has taken a broader, more community-focused approach. Its committee, including faculty, staff, and students, is charged with making endowment management recommendations to the board of trustees that embody Carleton's values. Recommendations can include the voting of proxy ballots, directly communicating with corporate management, writing shareholder resolutions, divestment, community investment, and investing in socially screened funds.[2] Yet another approach is the University of Vermont's Socially Responsible Investment Working Group, which includes trustees as well as faculty, staff, and students. The Working Group receives proposals from the university community regarding issues of social concern with respect to investments and shareholder resolutions. Successful proposals are then considered at the board level.[3]

While multiple models can and do work on campuses, it is critical to start this overall process by determining how the process will work most effectively within one's own institutional culture. The campus's political context is a wise place to start. A college or university with an activist student body who are passionately interested in purchasing green power and supporting local economic development will carry a direct impact both on the relative importance of the various topics and on how extensively the students will expect to be involved in the process. If the leadership team has a shared understanding of their goals for green investing and has been thoughtful about campus process, there is a greater likelihood of productive outcomes.

Independent of a college or university's approach to socially responsible investing and the process by which its community participates, from time to time a particular issue or corporation captures the imagination of a significant segment of higher education, and the issue of divestment is raised. In fact, the selling of stocks in companies that are seen to be acting irresponsibly has a history perhaps as long as the social responsibility movement itself. Although the most visible example to date has been apartheid in South Africa, there have been a number of other examples in the past half-century on such varied issues as stopping nuclear energy production, conducting business with totalitarian regimes like Burma or Sudan, and placing a halt on animal testing by cosmetic companies. When such an issue galvanizes a campus and the issue of divestment is raised, it can be a challenging moment for the board and the leadership team. Key to a reasonable outcome will be preexistent understandings among these groups about the college or university's mission, long-range goals, and best practices to resolve campus conflicts.

Getting Started: Four Strategies

For higher education institutions that have no previous experience with socially responsible investing and would like to begin carefully and gradually, there are a number of ways to approach investing through an environmentally responsible lens short of changing the institution's underlying investment decisions.

Proxy Voting

If the institution invests directly in securities, as opposed to through pooled funds, the active voting on shareholder resolutions has long been a starting point in both educating the campus community about key issues and introducing them to using a social-responsibility lens. In the months leading up to the annual meetings of publicly traded companies, shareholders may offer resolutions to be voted on at the annual meeting. Many of these resolutions address social responsibility issues, and with enough shareholder support, they can affect the behavior of the company even if the vote is well short of a majority.

A number of universities and colleges have had active proxy review committees for decades, although in retrospect interest on some campuses has waxed and waned depending on the current set of issues. Most of the activity of these committees has been devoted to reviewing shareholder resolutions that have been offered and determining how the institution's shares should be voted. A related approach, though less frequently used, is for the committee to develop proposed shareholder resolutions to be submitted by the institution. Using either approach is a good introduction to the complexities of making or supporting proposals that could result in requiring companies to change their behavior.

Direct Correspondence with Corporations

A second early step is to correspond directly with corporations. This is often an offshoot of the proxy review process and occurs when the institution wishes to support the substance of a shareholder resolution but finds the particular language of the resolution flawed. In addition, regardless of whether a college or university owns shares in a company, it can use its institutional position to advocate for or oppose a particular course of action by corresponding directly with the company. A number of institutions, including the University of Michigan and Smith College, have had extensive correspondence and discussion with the Coca-Cola Company in recent years as a result of the allegations made by activist groups about the company's anti-union and polluting practices in a number of other countries, including Colombia and India.

Student Demonstration Projects

A third approach is to create student demonstration projects. At Mount Holyoke College, when students developed a well-written and well-researched proposal to devote a small percentage of the endowment to socially responsible investing, the college's Investment Committee responded by assisting them in raising $25,000 and requesting that the students run a pilot program of socially responsible investing to demonstrate its efficacy. The student group, with support from the Development Office, a professor of economics, and the assistant treasurer, has developed the asset allocation, made investment decisions, and reported fund performance annually to the Investment Committee. Their investments have included both community development and socially responsible mutual funds. The demonstration project is now four years old and has actively and productively engaged the group of students most committed to the issue. The Investment Committee has periodic discussions with the students about their results and whether there is a role for socially responsible investing within the endowment itself.

Direct Community Development

Another approach is direct community development. The rationale for community development as a component of environmental sustainability is twofold. First, support for local production of goods and services makes regions more self-sufficient, reducing energy used in transportation, packaging, and so forth, and using local resources more wisely. Second, lessening poverty, providing jobs, and promoting environmentally responsible economic development are necessary components of a sustainable future. A number of colleges and universities have been actively engaged in supporting economic development both directly and in partnership with their local communities. Many institutions, including Yale University and the University of Pennsylvania have led significant neighborhood development efforts in the areas surrounding their campuses.[4] A survey of fifteen selective liberal arts colleges revealed that eight of them are currently making direct community development investments.[5] The Ball State University Foundation invests in venture capital and mezzanine loan funds with a concentration of Indiana investments, in the areas of health care, technology, and housing.[6] In addition to direct financial support, colleges and universities support local economic development in other ways, from purchasing food from local growers to giving priority to regional contractors.

Modifying Endowment Investments

For schools that wish to take a more comprehensive approach to green investing, the asset allocation of the endowment can be altered to include an explicit focus on environmental stewardship. This typically results either in a portion of the endowment being targeted to investments that further the institution's commitment to environmental sustainability or to the imposition of an environmental screen. In practice, institutions that employ a socially responsible investment philosophy will often use a combination of these two approaches. The easier of the two to implement is targeting a portion of the endowment to particular kinds of investments, since this approach can be incremental and does not impact the balance of the institution's investment decisions. A growing number of funds, both publicly traded and privately held, focus on green energy and other environmentally responsible investments. In addition, useful indexes that track green stocks are emerging, including the WilderHill Clean Energy Index and the NASDAQ Clean Edge Green Energy Index.

A second option is direct investment in companies that offer products or services that are compatible with the institution's environmental goals, either through individually managed portfolios or, for larger institutions with in-house investment expertise, by making direct investments in the companies themselves. At the University of Minnesota, for example, the endowment invests in private funds that develop real estate and buildings to LEED certification requirements and with private equity managers who invest in clean technologies and alternative energy companies.[7] This approach also need not be confined simply to equity investments. One possibility that has the added advantage of fostering regional economic development is to provide loan capital to local organizations that offer financial services to low-income individuals and small business start-ups. Examples of such organizations are the Cooperative Fund of New England in western Massachusetts and the Self-Help Credit Union in Durham, North Carolina.

A more comprehensive alternative is to employ some type of environmental screen to all institutional investments. Such a screen would favor green or neutral investments and reduce or eliminate investments that are judged to be at odds with the institution's environmental goals. This is a more radical approach to green investing that requires more extensive research and a carefully crafted policy on screening as well as the expertise to apply such a screen while continuing to produce adequate investment returns. In light of these challenges, socially responsible screening approaches to date have focused mostly on relatively narrow issues like investment in Sudan or green energy production, rather than trying to develop a comprehensive matrix of companies or funds to favor or to avoid.

A different model, which continues to grow in practice, is to actively seek investments that carry both a competitive return and positive environmental impact. Yale University's 2009 report on the endowment describes a number of green companies that are part of the university's venture capital investments.[8] This approach is likely to become the most common way in which environmentally responsible investing is enacted in the future. Still, before deciding how far to go down the path of incorporating environmental sustainability into an institution's investment process, there are some important caveats to consider.

Four Investing Cautions

Mission Compatibility

Virtually all colleges and universities are actively considering how to be more environmentally responsible and how to incorporate other aspects of sustainability into their operations. However, green investing, because of the uncertainty of the impact on endowment return, has been much slower to develop, even in small ways. Higher education institutions seriously considering green investing need to consider carefully how it accords with their missions.

The Role of the Endowment

Does the college depend on investments for a significant portion of its operating budget? If so, any sacrifice of return that might occur through limiting the universe of available investments will have a significant negative impact on the institution. While the common wisdom has been that any limitation of investment opportunities risks lower returns, that point of view is evolving. A survey of investment consultants conducted by the Social Investment Forum Foundation in September 2009 reports that two-thirds of the consulting firms surveyed offer advisory services in one or more areas of socially responsible investing. A majority of respondents believed that a range of strategies including asset selection according to sustainable themes, corporate engagement—including a social responsibility analysis into investment decision, proxy voting, and the inclusion of socially responsible stocks and bonds in a portfolio—has either a positive or neutral effect on performance. However, on the decision to exclude stocks and bonds from investment portfolios, the response was decidedly negative, with only 8 percent responding that this approach resulted in a positive impact on performance.[9] As the body of research and analysis grows and matures, endowment managers will achieve a better understanding of the interrelationship between screening and investment returns.

The Trend toward Pooled Funds

A growing complication to sustainability-driven investing, even if the endowment's asset allocation is not changed, is the growing trend toward investing largely or completely in pooled funds rather than in individual portfolios of securities. To the extent that a university or college invests in pooled funds, there is no ability to influence how the fund is invested, and the only level of control is to invest or not to invest in the fund. Taken to its logical extreme, an institution determined to use an environmental screen for all of its investments would be limited either to funds with an environmental focus, which remains only a small percentage of available investment vehicles, or to the more expensive alternative of individual, actively managed funds. Universities with large endowments that typically retain significant in-house investment expertise are in the best position to enact this strategy fully.

The Impact of Institutional Philosophy

The final consideration is a philosophical one: Should anything other than probable return influence how the endowment is invested? There are, of course, a wide range of opinions on this issue. Some will argue that, regardless of its size, an institution's endowment is a key asset that should not be constrained by considering anything but probable return at a reasonable level of risk with the understanding that environmentally responsible companies or funds can be compatible with this point of view. Others may contend that using investment decisions to encourage environmentally responsible behavior is ethically the right approach and that some reduction of investment return, if that is the result, is no different than supporting the additional cost of green buildings or green energy development.

At the core of an institution's determination to consider social responsibility issues in investment decisions is the tradeoff between social and monetary benefits. Kyle Johnson of Cambridge Associates describes this challenge succinctly: "If the practice of social investing comes at some monetary cost, the key question becomes whether the incremental social returns derived from incurring that cost outweigh the social returns that would otherwise have been generated from simply spending the greater monetary returns earned through 'traditional' investments."[10] Before embarking on a course of action that will limit investment choices, colleges and universities need to address this question and reach a thoughtful conclusion as a community.

Three Emerging Trends

- Growing interest in socially responsible investing, particularly green investing
- Growing support for the United Nations Principles for Responsible Investment
- Growing focus on incorporating green investing as a component of institutional sustainability rankings

Three Emerging Trends

After decades of playing a relatively marginal role, socially responsible investing is poised on the brink of becoming a significant factor in making investment decisions in higher education. The financial services industry is paying increased attention to this model of investing, and the investment options are growing. The Social Investment Forum, a national trade association for the sustainable and environmentally responsible investing industry, produces a biennial report on Socially Responsible Investing Trends in the United States. The 2007 report, released in early 2010, describes a rapidly growing industry. At the time of this report, 11 percent of assets under professional management in the United States were involved in socially responsible investing, including screening, shareholder advocacy, and community investing, representing 2.71 trillion dollars. The report goes on to describe an industry reacting quickly to the growing interest in sustainability-driven and environmentally responsible investing.[11] The Social Investment Forum's 2009 report on Investment Consultants and Responsible Investing updates these findings, reporting near-universal agreement among the consultants surveyed that interest in socially responsible investing will persist and expand.[12]

Global interest in socially responsible investing is also growing. In 2005, former United Nations Secretary General Kofi Annan initiated the United Nations Principles for Responsible Investment (UN-PRI). These six principles include the following: incorporating environmental, social, and governance (ESG) issues into investment analysis and decision-making processes; being active owners and incorporating ESG issues into ownership policies and practices; seeking appropriate disclosures on ESG issues from the entities being invested in; promoting acceptance and implementation of the Principles within the investment industry;

A Decision One Would Make Differently with Hindsight

- Develop an integrated institutional approach to environmentally responsible investing rather than addressing issues individually through a piecemeal approach.

working together to enhance effectiveness in implementing the Principles; and reporting on activities and progress toward implementing the Principles.[13] The UN-PRI website identifies 747 institutional investors world-wide who have endorsed the principles as of June 2010, 109 of whom are from the United States.[14] The early adopters include primarily investment managers and consultants. The twenty asset owners who are currently represented are dominated by pension funds together with a small number of foundations. No colleges or universities have yet become signatories.

A third trend is the addition of environmentally responsible investing into national rankings of college and university sustainability efforts. The most prominent is the College Sustainability Report Card, which conducts an annual comparative assessment of sustainability practices in a number of more traditional areas, including climate change and energy, food and recycling, and green building. Significantly, however, the Report Card also includes ratings of endowment transparency, investment priorities, and shareholder engagement.[15]

While it is possible to quarrel with the particular areas of emphasis and the approach to measurement practiced by this or any survey, the most significant aspect of this particular survey is the prominent place it assigns to green investing. This is viewed as a harbinger of a future in which environmentally responsible investing is assessed as central to college and university sustainability efforts. When this happens, expectations will rise, both on and off campus, for endowment management to become an even larger component of an institution's approach to sustainability.

Conclusion: Achieving Sustainable Financial Goals

The history of socially responsible investing in higher education over the past several decades has been cyclical, with a very few periods of intense interest, such as during the height of the South Africa divestment push. However, in recent years, sustainability and environmental concerns have formed an expanding element of the college and university financial sector.

While the vast majority of endowment investments in higher education are still made using a strictly financial lens, there is growing interest in and attention to sustainability as a key consideration in the investment decisions being made by colleges and universities. While relatively few institutions are approaching this in the more traditional manner of social screening or allocating fixed percentages of the endowment for sustainability and socially responsible purposes, there are more and more examples of institutions that are seeking investments in green companies and technologies, in renewable resources like timber, and in regional economic development opportunities. Additionally, the notion of integrating an understanding of a particular investment's environmental impact into the overall assessment of that investment's value for the endowment is leading to a much more systematic inclusion of sustainability in investment portfolios.

As universities and colleges continue to evaluate the role of environmental responsibility in investing their endowments, it is particularly important to be clear about what the institution is trying to accomplish and to probe the assumptions behind their decisions. If sustainability is important to fulfilling the institution's mission, how can this best be accomplished? If sustainability-driven investing is one aspect of the institution's approach, what are the expected outcomes? Will investors increasingly seek environmental sustainability as a desirable aspect of corporate policies and therefore expand investments in these areas? If there is a lack of sufficient capital devoted to sustainability initiatives, does this result from an expectation of below-market returns? Is the institution willing to accept lower returns on investments with a sustainability focus and commitment? Careful evaluation of these and similar questions will provide guidance as institutions consider their options in environmentally responsible investing. While still not seen as a core component of sustainability thinking on some campuses, this

Single Most Important Piece of Advice for a New Professional

- Proceed deliberately and consult with colleagues at similarly situated institutions about developing a program of green investing before committing any resources.

A Sustainability Myth

- There is an inevitable tension between being environmentally responsible and being cost effective. In investing, as well as in the broader environmental stewardship arena, if it does not make good economic sense at some level, then it is unlikely to be sustained.

template for investing is one more aspect through which an institution's commitment to sustainability can be achieved.

As interest in environmentally responsible investing grows, as the approaches to doing so become more varied and sophisticated, and as companies providing green products and technologies become more competitive and produce greater investment returns, this element of sustainability will become more influential and mission-centric over the next two decades of leadership. The current moment is a propitious one for colleges and universities that have not already done so to consider the role of environmentally responsible investing in the institution's investment decisions and sustainability policies and programs.

CHAPTER THIRTEEN

Sustainability and Higher Education Architecture: Best Practices for Institutional Leaders

Scott Carlson

Green Building: Past, Present, Future

New buildings can transform campuses, as any savvy administrator knows. In the realm of green building, transformative campus buildings abound, and the Park Center for Business and Sustainable Enterprise at Ithaca College stands as one of the best examples.

In 2002, the college—a solid institution with little to distinguish it from other liberal arts colleges in the Northeast—was planning to construct a new business school building to help revitalize its business program. Then a buzz started: People at the college started talking about making the building green, *really* green. They batted around the idea of getting a platinum rating in Leadership in Energy and Environmental Design (LEED), the popular green building rating program. At the time, many of the institutional buildings that had gotten LEED platinum certification had been tied to environmental programs. Administrators at Ithaca College thought that constructing a highly rated green building for the business school would be the "man bites dog" hook that would attract attention and maybe even set a precedent for green building and sustainable practices on campus.

These ideas became a reality when Dorothy D. Park, a donor to the college, gave seven million dollars for the building on the condition that it would strive for LEED platinum. When administrators at Ithaca College courted her, they showed her *The Next Industrial Revolution,* a film that documents William McDonough's work on the Adam Joseph Lewis Center at Oberlin College, one of the earliest, most ambitious, and best-known green buildings in academe. According to a former provost at Ithaca College, who had been an advocate for a green build-

Figure 13.1 The Dorothy D. and Roy Park Center for Business and Sustainable Enterprise at Ithaca College. This building helped rebrand Ithaca College as an institution with a reputation for sustainability. Courtesy of Ithaca College. Photo by Adam Baker.

ing, Park made a prediction after the film was over about Ithaca's green business school: "This building will put that little college on the map, won't it?"[1]

The Park Center for Business and Sustainable Enterprise was indeed transformative. As the building went up, Ithaca College built a reputation as an institution with sustainability values and high-profile green efforts like signing the American College & University Presidents' Climate Commitment (ACUPCC). The building also set a precedent on campus: in 2009, the college opened another building designed to meet LEED platinum standards—an administration building even more ambitious than the Park Center, with an innovative natural ventilation system and other green features. In 2008, the Association for the Advancement of Sustainability in Higher Education (AASHE) recognized Ithaca College as a sustainability leader among peers, in part for the college's quest to become the first institution to have two LEED platinum buildings. Both buildings cost around $20 million, with a modest 3–5 percent premium for their green features, which will be paid off within five to seven years through energy savings.

Figure 13.2 Occupants of the Peggy Ryan Williams Center, the second LEED platinum building on the Ithaca College campus, may have to adjust their behaviors to help the building's natural ventilation system meet its energy efficiency goals. Courtesy of Ithaca College. Photo by Bill Truslow.

In the beginning, Ithaca College administrators were told that striving for LEED platinum could cost the college some 15 percent more than conventional construction. But while the projects had their share of challenges—like finding architects and construction companies with LEED experience—good management strategies kept costs down. When people ask Rick Couture, vice president of administrative services at Ithaca College, what they should know about building ambitiously green, he has a stock reply: "It is not as daunting a challenge as it is made out to be."[2]

Ithaca College is just one of scores of colleges and universities that have joined the green building boom of the past decade. In a way, green building is as old as construction itself. For millennia, people have designed buildings that drew materials from their immediate surroundings and interacted with the local climate. Consider the dense stone or mud walls of traditional desert architecture like adobe buildings; they absorb heat and keep the interior of a building cool. Traditional features like operable windows and cupolas have been used both to vent heat and provide light.

LEED Certification: Is It Right for Your Institution?

That we now distinguish some buildings as "green" gives an indication of how far we have strayed from these traditional practices during the twentieth century, when cheap energy could transport raw materials around the globe and light, heat, or cool any interior space. The postwar architecture fashions of Modernism—and especially its subgenre, Brutalism—led to buildings with monolithic walls with no windows, or curtain walls made entirely of windows, none of them operable. In part because of pervasive design decisions like these, the built environment in America sucks down a tremendous amount of energy and spews out a proportionate amount of pollutants and waste. According to some commonly cited statistics, buildings consume about 70 percent of the nation's electricity, emit nearly 40 percent of the nation's greenhouse gases, and generate 70 percent of landfill waste.[3]

But a movement in green design had begun to take hold in the late twentieth century, with pioneering architects like Malcolm Wells, William McDonough, Mike Reynolds, Sim Van der Ryn, and others. The Leadership in Energy and Environmental Design Program was started in 1998 by the U.S. Green Building Council, a nonprofit devoted to promoting environmentally sustainable building. Over the past decade, the program has developed widespread acceptance in academe as a leading standard for green building certification, a way to display sustainability credentials in something akin to Olympic-medal colors: silver, gold, and

Living in the Unity House

MITCHELL AND CYNTHIA THOMASHOW

We live in a glass house on a college campus. If that is not enough to catch your breath, we live that way on purpose, with full disclosure about our use of energy, water, materials, and the waste stream.

When we arrived at Unity College in central Maine, we came with the intention of developing an exemplary sustainable campus. We decided that the best way to model a sustainable lifestyle would be to live according to our own values. Our first step was to convince the board of trustees that Unity College had an extraordinary opportunity to build a reasonably priced LEED platinum president's residence on campus. In partnership with Bensonwood Homes and the MIT School of Architecture, we built the Unity House, a zero-carbon, light footprint, energy efficient home.

We anticipated that this residence would be an immediate and high-profile way to educate the campus and the general public about living a sustainable life, but what did this mean for us more personally? We were about to find out. People now drive up the road to the college and invariably pause to stare at the stunning array of thirty solar panels on our roof. Nestled at the edge of a bucolic pasture, the modern design is a provocative surprise. The panels glow a silvery shade of blue on a sunny day, drawing unexpected attention. I have to remember to wear suitable clothes at all times just in case they get out of their

Unity House. Courtesy of Unity College, Unity, Maine.

cars to walk around. Spontaneous tours are becoming my specialty. This creates an interesting tension. We value our privacy, yet we follow our urge to use this space as a teaching laboratory. Balancing this tension is now a daily practice. In many respects, this challenge models the essence of a college presidency. One lives a public life. In the case of Unity House, we feel compelled to invite people to visit, simply because we are enthusiastic about the importance of living a sustainable way of life.

Unity House was designed to perform both as a private home and a public venue. We now host board meetings, small retreats, and mini-conferences. A gathering room in the center of the house comfortably holds about fifty people. The wall at the west end of the room folds up to give us an additional 270 square feet. The wings of the house serve as a private domain. We ensure privacy by sliding tall wooden doors into place. Nevertheless, we continually open our private quarters to visitors because people are genuinely interested in how we live, and we are anxious to share those aspects of the experience as well.

Buildings are powerful teachers. However, we rarely ask deep questions about the buildings we inhabit. We learn good and bad ecological habits from them. They influence the way we relate to and respond to the places where we live. We believe that buildings should be transparent—constructed and interpreted so that the dweller is always reminded of basic ecological principles—where energy, food, water, and materials come from. Ideally, buildings link indoor experiences to a deeper understanding of place and landscape. When we explain to visitors that all of the materials in Unity House come from New England or are recycled, they are impressed because they can viscerally trace the flow of materials from their origin to their location in the house.

Consider the different ways the house has been utilized in its first year of existence: A dozen sections of various Unity classes have visited the house to learn about passive solar design, active solar collection, ecological landscaping, LEED certification, the use of recycled materials, zero waste construction, and modular building construction.

- One class is using the house to develop an environmental education exhibit about its origin, special features, and educational impact.
- We host open houses for parents and students explaining the principles and applications of sustainable living.
- We are now a destination for regional home tours of ecological residences.
- The house was the main venue for a small conference on the Art of Stewardship.
- We routinely host dinners for potential donors who are interested in sustainability.

(*cont'd*)

Living in the Unity House

As the first North American college president's residence to receive a LEED platinum rating, Unity House has been covered in journals, on television, by newspapers, and on websites. People often ask if we tire of living in such a public arena. Our glass house raises the stakes of our community work. It reflects not only our abiding interest in energy efficiency and sustainable design but also the importance of transparency and collegiality. The Unity House is an educational experiment. Our experience of living in the house is less about the dwelling per se and more about achieving a broader vision. We hope to inspire Unity College's institutional excellence as a leader in sustainability and environmental studies. We realize that we are stewards of the home and pioneer dwellers on a campus that will serve as a sustainability laboratory, and we believe that if this can be accomplished in remote, frugal, sun-deprived Maine, it can be accomplished almost anywhere.

Mitchell Thomashow has been president of Unity College since 2006. He is the author of Ecological Identity: Becoming a Reflective Environmentalist *(1995) and* Bringing the Biosphere Home *(2001). Cynthia Thomashow is Executive Director of the Center for Environmental Education and a faculty member at Unity College.*

platinum. (The lowest, basic certification was once known as "bronze," but the U.S. Green Building Council dropped that medal name because people associated it with third place, and no one wanted to come in third.) Eight higher education projects were certified or registered with LEED in 2000, 107 in 2002, 336 in 2004, 779 in 2006, and 1,534 in 2008.[4] Officials at the U.S. Green Building Council expect the number of LEED-registered or LEED-certified projects at colleges to exceed the number of colleges themselves in 2010.[5] With this rate of adoption, LEED is fulfilling one of its primary goals: to mainstream green building materials and green building techniques.

However, the program has also generated controversy—in part because of the very way that the program was designed to appeal to conventional builders and designers. LEED is based on a point system, and earlier versions of LEED included easy points that could help make a certification attainable for builders who wouldn't normally aspire to environmental features. In an earlier version of LEED, including a bike rack outside a building and changing room for cyclist commuters, for example, would earn one point—the same credit a builder would get for preserving 75 percent of an existing structure. Critics charged that giveaways like these led to "point chasing" that would make a LEED-certified building no more green than a conventional building.[6] Walter Simpson, the former energy officer at the University at Buffalo, once described a point that a UB building got for having an electric car recharging station in front of a building. "There are no electric cars here," he said. "The charger has never been used in the five years it's been there. It was a cheap point to get."

College administrators also grouse about the cost of LEED certification, which can be tens or hundreds of thousands of dollars. (Some architects estimate certification can cost around $2,000 per point.) Certification includes third-party commissioning of the building—that is, testing whether the energy-saving materials, devices, and designs are actually saving as much energy as they are supposed to.[7]

Some colleges have skipped LEED certification to build "LEED equivalent" buildings or to follow other green building standards in order to save money, avoid hassles, or emphasize their priorities. In building a $400-million engineering quadrangle, Stanford University devised its own rating system based on LEED and Labs21, a green building program for science buildings.[8] Douglas S. Kelbaugh, the former dean of the A. Alfred Taubman College of Architecture + Urban Planning at the University of Michigan, persuaded administrators not to get LEED certification for an addition to the architecture school building, arguing that "it makes more sense to spend that $100,000 on photovoltaics or better windows or insulation." Kelbaugh complained that LEED did not grant the project credit for building on top of the existing structure, thereby saving land. Kelbaugh also said that another green building standard—the American Institute of Architects' 2030 Challenge, which he chose to follow—had more emphasis on reducing the carbon footprint than did LEED.[9] (The 2030 Challenge calls for new buildings to reduce energy use by 60 percent by 2010, 70 percent by 2015, 80 percent by 2020, and 90 percent by 2025, and be carbon neutral by 2030.)

The U.S. Green Building Council responded to the criticisms with a new version of LEED, released in 2009. The new system puts more emphasis on credits that address the most important environmental issues, like climate change and fossil-fuel depletion. The new LEED system also allowed architects and builders to earn bonus credits for regionally important designs and systems—for example, water-saving techniques in desert climates or wetlands protections in riparian zones. The U.S. Green Building Council also launched the "Portfolio Program," which is designed to help some builders save money on the certification process.

Those that plan to certify multiple structures under LEED can join the Portfolio Program to get certification for policies and practices like the use of green products in cleaning materials that would be applied to every structure built.[10]

Programs like these may make LEED certification more attractive. But advocates for the LEED program have long argued that the clamor over the cost of certification is shortsighted and overblown. Robert Koester, a professor of architecture and director of the Center for Energy Research/Education/Service at Ball State University and coauthor of chapter 8 in this volume, is one. A certification cost of $100,000 might sound significant to a chief financial officer who is watching every dollar on a project, but that amount is nothing compared to the overall cost of the building, he argues. "The problem is that if we don't measure we can't prove that we're doing what we set out to do," he says. Commissioning a building has value in that it provides quality control on the building process. "By avoiding [certification] expenses, you wind up shortchanging what is possible to achieve."[11]

Koester adds that an institution saying that it is building to an "LEED equivalent" without certification or commissioning runs counter to our certification practices in other industries. "What if we said to the automobile industry, We want you to build to this standard, but don't bother testing anything—just tell us you're committed and we'll believe you," he says. "Universities themselves are used to accreditation. They benefit by showing that they set out to do what they said that they were going to do." The LEED brand, which is widely recognized even by people who are not in the building industry, also carries significant influence in conveying commitment to sustainability values. The 2030 Challenge, by contrast, is not nearly as well known to the general public, whatever its merits. Stanford's own rating system for its engineering quad means little to anyone outside of Stanford's administration.

Beyond Certification: Best Practices for Higher Education Leaders

Whether colleges build to meet LEED standards, or another standard, or even their own, their green building practices should follow some standard best practices, which Koester elucidates.

First, set goals and follow an integrated design process from the very beginning of the project. That means gathering everyone who will be involved in the design, construction, and occupancy of the building, including architects, engineers, construction managers, and consultants on lighting, acoustics, interior design, landscape, or technology. Of course, that group should also include representatives from the college: the owners or occupants and administrators from finance and facilities.

In their meetings, these people need to learn to talk to one another and start to form a collective vision about the shape and goals of the project. *Design Intelligence,* a trade publication for the design industry, noted that the integrated approach was gaining popularity and vital to the growth and success of the industry. Those who followed the approach had to check their egos at the door and "embrace the idea of collaboration."[12]

"The wisdom of the group is far greater than any individual," Koester says. "You can make enormous strides early in the process, such that when you get toward the end of the design development and construction document work, there is little left to do. The construction documents just fall into place, because so many decisions have been made and so much is understood from the early days of the project."[13] This is not how projects traditionally begin. Koester likens the traditional process to boxcars on a train: An architecture firm is hired and produces a conceptual design; then an engineering firm reviews the design; then various consultants add to the design; then a construction firm signs on; and so on until at the end the grounds are spruced up by a landscape crew.

In green design in particular, this traditional boxcar approach hurts the final product. First, the best green buildings operate through complex, interdependent systems. The window choices may increase the heat in the building, which affects decisions on HVAC systems, which might affect decisions for solar power systems, which might have implications for the design of the roof, and so on. (Landscape architecture, which is too often an afterthought in conventional building, can make significant contributions to important design decisions like heat gain or storm water management.) A process that produces a design, then hangs "green" elements on that design, may not be as successful as a more integrated process.

For budget-conscious administrators, there is another compelling reason to follow an integrated design process: it costs less. Koester says that colleges have traditionally adopted the boxcar model because they do not want to have to pay firms and consultants until they absolutely have to. However, following the train metaphor, if the design has "left the station" with the architects, the engineers and consultants who follow may have to work especially hard to make their recommendations conform to the shape of the already-conceived building. "The irony is that it ends up making things so complicated later that there really is no net savings," Koester concludes.[14]

The question of added investment in green features is always a prominent, much-discussed issue in green building. In a 2006 study, Davis Langdon, a construction consulting firm, found "no significant difference in average costs for green buildings as compared to non-green buildings" after analyzing more than 220 construction projects, 83 of which were going for some level of LEED certification.[15] Many architects say that there is no premium for LEED silver or even some LEED gold projects.

When green building costs more, many colleges and universities frame the added investment in terms of years until payback: How long before energy savings pay for the up-front investment? As Koester points out, "years until payback" is not a metric we apply to other significant purchases. We do not consider payback when investing in, say, a car. With these purchases, we usually make choices on the basis of quality—how a car drives, its reputation for reliability, or its sleek look. The payback mindset "pessimizes" the conversation, Koester adds, making something like a ten-year payback seem like a long time. Instead, colleges could think of green features in terms of return on investment. A ten-year payback could be seen as a guaranteed 10 percent return—a respectable rate by any market standard.[16]

How Green Buildings Return a University's Investment

Return on investment in green building comes in many forms, some difficult to quantify. A look at a project at Northern Arizona University provides a sense of costs and benefits of green building. Northern Arizona University constructed the highest-rated LEED building in academe—a lab building, a notoriously inefficient building type. In 2001, John D. Hager became president at Northern Arizona University and set out to remake and strengthen the institution through research and new construction. (As the state's old teachers college, the university had long been in the shadow of Arizona State University and the University of Arizona.) An environmental ethic had been gaining momentum on the campus, says Richard Bowen, associate vice president for economic development and sustainability. Bowen was part of a group on campus pressuring the president to commit to building green, starting with the new Applied Research and Development building, known to most on campus simply as ARD.[17]

The university hired Michael Hopkins, a prominent British architect, and started shooting for the greenest building possible. "We wanted to drive this whole green-building idea," Bowen says. "We thought we could push the envelope." The design and construction of the building included some remarkable innovations and extra efforts, particularly in its concrete and in its air handling. The concrete uses a high percentage of fly ash, a byproduct of coal-burning power plants that can be substituted for Portland cement to reduce concrete's carbon footprint and make it stronger when cured. But fly ash retards the curing process—it

Figure 13.3 The Applied Research and Development building at Northern Arizona University went over budget in the design stage, but energy efficiency measures will allow the university to recover those costs within ten years. The innovative design of the building has brought significant international attention to the university. Photo credit: Amanda Voisard.

has a 56-day curing cycle, versus a 28-day curing cycle with conventional concrete. The building's foundations were poured in summer and fall, but the vertical columns had to start going up in the middle of winter when it was 15 degrees outside. If the columns froze, the pour would be ruined. So workers went to big-box stores around Flagstaff, buying every electric blanket available, then used the blankets to wrap the columns for up to a week—enough time to allow the concrete to set up.[18]

Like most laboratory buildings, the ARD building must flush air out of labs to prevent contamination. In most buildings this is accomplished through high-velocity fans, which make most lab buildings highly inefficient because they constantly blow out conditioned air. Designers of the ARD building found that they could move air more efficiently and more effectively by using big fans that turn very slowly. (High-velocity fans create eddies throughout buildings.) A heat-recovery system captures energy that would normally be lost through this ventilation. Designers and officials at Northern Arizona University had to get special permission from regulators to install these systems. But the work this involved was worthwhile. These innovations, along with passive solar design, contribute to the building's remarkable efficiency—about 80 percent more efficient than a conventional building of its type.[19]

But innovations like these came with costs, financial and political. At $24 million, the building ran about 10 percent over its projected budget, mainly in the design stage. President Hager was worried. "He said, 'If this goes sour, your heads are on the block,'" Bowen recalls. "That was the agreement." State officials had taken notice of the cost overruns too. "I was tossed out of the legislature several times over sustainability because people thought it was the domain of Birkenstock-wearing hippies. A couple legislators called it 'that damn hothouse building.' That just tells you how ignorant they were."

Northern Arizona University got better at explaining sustainability and green building to policy makers, Bowen says, "but we stopped putting it in environmental terms and started putting it into financial terms." Now he can rattle off the benefits ARD brings to the university and Arizona taxpayers: the building, which is designed to last a hundred years, has an estimated annual energy savings of $250,000, which will pay for the cost overruns within ten years. The university has since built more LEED buildings and has saved $1 million to $2 million in design and construction costs by using what it learned from building ARD. The innovations in the concrete and air handling led to university-owned patents; the university allows Arizona architects and construction companies to use this intellectual property, giving the companies a competitive advantage that leads to an economic impact on the state that is "somewhere in the millions," Bowen estimates. The building has also raised the profile of the university: The project has won several awards, including a 2009 International Award from the Royal Institute of British Architects, the only project in the United States to do so that year. The Arizona Public Service, the large utility in the state, gave the university a $1 million gift because it wanted its name associated with the building. Representatives from major companies such as Walmart have asked to tour the facility to learn more about sustainability.[20]

In the end, "those same state senators who wanted my head for a while came to the building dedication and took all the credit because it became cool," Bowen says.

The Greenest Building of All: None

There is an adage among architects, planners, and sustainability advocates: The greenest building is the one never built. This chapter on green building would be incomplete if it did not present the possibility of not building at all. That possibility might sound zany to college administrators who have been obsessed with building and growth over the past decade, but it might be one of the most sensible approaches to college planning for the coming decades, given the challenges that colleges may face. "Sustainability" is not just about solar panels, graywater systems, recycled materials, and green roofs; college administrators should question whether their pace of building—green or not—is economically sustainable as well.

Building has been at a breakneck pace in the past decade. According to Sightlines, a company that gathers facilities data from a diverse set of about two hundred institutions, an average of 14 percent of space on those campuses was built between 1998 and 2008. That 14 percent represents more than 61 million gross square feet. Among research institutions, the average proportion of new space was higher—17 percent built in those ten years.[21] Campus planners have noted that the growth of colleges has far outpaced growth in students and faculty. Reliable data are difficult to come by, but campus-planning experts like Philip Parsons of the Boston architectural and planning firm Sasaki have made estimates. Cobbling together figures from various sources, Parsons calculates that higher education institutions had 1.3 billion assignable square feet in 1974 and a full-time-equivalent enrollment of 7.8 million students, or about 160 square feet per student. Today the assignable square feet at colleges is estimated at about 6 billion square feet and growing, and there are about 13.2 million full-time students, or about 450 square feet per student. "The space per student has in some cases tripled since the 1970s, even while the size of the American home doubled," Parsons says. "Colleges have been prodigal."[22]

Many chief academic officers would rightly point out that spaces on their campuses have grown to respond to different teaching styles and new kinds of research. The packed lecture hall, with chairs and desks close together and bolted to the floor, is not considered an optimal modern learning environment. However, the "arms race" in campus amenities has also been a major driving factor. Parsons says that the primary growth in campuses has been in student centers, recreational centers, residential space, sports facilities, and libraries (whose space has grown even as collections are pushed to off-site storage facilities). One can generalize by saying that the colleges with the biggest endowments generally have the most space per person on campus. In some cases, the growth is driven by student fees that are higher than tuition, "which means that we are putting money into buildings that have nothing to do with the core educational mission—because we can," Parsons says.

Colleges and universities cannot do this much longer. Many higher education administrators do not realize—or choose to ignore—the fact that the bulk of the total cost of a building comes after it has been built. Some 70 percent of the life-cycle costs of a building are in its operations, maintenance, utilities, upgrades, and renovations.[23]

In an alarming number of cases, colleges cannot support the girth of what they have already built. The deferred maintenance backlog at the University of Maryland at College Park, for example, is more than $600 million.[24] The Kansas Board of Regents has estimated that the maintenance backlog at Kansas's six state universities and various two-year colleges amounts to about $1 billion.[25] An administrator at the University at Buffalo has estimated the university's three campuses need repairs and upgrades amounting to about $1.5 billion.[26] Colleges that cannot support their existing buildings, while continuing to build new ones, risk operating in what people in the facilities industry call "run-to-failure mode"—that is, running buildings to systemic failures by not keeping up with vital repairs.

Moreover, maintenance budgets have not kept pace with the increasing complexity of building systems. Newer systems—especially those in green buildings—can be more expensive and more difficult to repair than older ones. Those systems need more frequent attention to make sure they are achieving the kind of energy efficiency they were designed to get. This kind of regular maintenance may carry great weight in years to come. Officials from the U.S. Green Building Council have said that underperforming green buildings may lose their silver, gold, or platinum plaques under future versions of the LEED program.[27] Yet deferred maintenance is only one problem. If the United States continues to experience a steady climb in energy prices, the lighting, heating, and cooling of millions of gross square feet will be a major burden for every higher education institution.

As one answer to this problem, more planners and sustainability advocates have presented a solution that sounds simple yet is politically dicey and difficult to pull off: stop growing. "No net growth" is one term for the concept. If the college needs to build something new, an equal amount of square footage has to come down. Since colleges are generally reluctant to remove buildings from their campuses, such a policy would likely favor renovations of existing space, which can be a greener alternative to demolition and new construction. In the summer of 2010, Ohio State University adopted a no-net-new-academic-space policy; administrators said they merely needed to improve the sufficient space they already had.[28] Parsons, of Sasaki, who had helped Ohio State come up with its plan, is trying to persuade other clients to adopt a no-net-growth policy.[29]

Even if such policies are not adopted, simply better utilization of space would be a more sustainable, more practical approach to dealing with growth. At Kean University in New Jersey, classroom utilization on Fridays was 11 percent and 8 percent on Saturdays. This was financially and environmentally unsustainable when the college was spending $16 million a year to heat and cool buildings seven days a week. In 2009 Dawood Farahi, the president of Kean University, offered students discounts of up to 20 percent for taking classes on Fridays and Saturdays. The university was able to enroll 700 more students, which meant that Kean could raise tuition by only 5 percent instead of 20 percent after state budget cuts.[30]

One of the most confounding issues for administrators seeking space efficiencies is faculty and staff offices, which can make up 25 percent of space on a research university campus, compared to the 5 percent that is classrooms. Unfortunately, administrators have a hard time changing attitudes about campus office space, which is perceived as a right no matter how little it may be used.

Conclusion: The Future of Green Building

Growth will likely continue on college campuses because new buildings are among the things administrators point to as evidence of achievement. A crucial question is: What kind of buildings will colleges put up in the future? In the summer of 2008, Princeton University opened a new library designed by Frank Gehry. The university spent at least $74 million for 87,000 square feet, or more than $850 a square foot.[31] Princeton was not just paying for custom steel beams or 88,000 pounds of embossed stainless-steel panels imported from Sweden. It was also buying the Gehry brand. Especially over the past two decades, colleges have been caught up in erecting buildings by star architects, bringing media attention and alumni admiration.[32]

The "starchitecture" of the future will be different. A star building will be judged not only by its looks but also by its environmental impact and energy efficiency. Already, policy makers have been discussing concepts like net-zero-energy buildings, or buildings that produce as much or more energy than they use. The American Clean Energy and Security Act of 2009, known for its cap-and-trade requirements, tried to establish energy codes for buildings to reduce their energy use by 75 percent by 2030. Deval Patrick, governor of Massachusetts, formed a task force of architects, developers, environmentalists, and energy experts to study the net-zero concept. The task force's report, released in 2009, laid out a plan for making net zero mainstream in the state by 2020 and mandatory by 2030.[33]

College administrators may start to look at their campuses as an organism in which the "waste" of one building can be the "food" for another. That is the idea behind a $70 million geothermal project at Ball State University, which links the university buildings to some 3,700 geothermal wells on the campus. The university will draw heat from the ground for its buildings in the winter and store heat there in the summer. But the project will first establish a balance between the energy use in the various buildings. If, say, one building is overheated, that heat could be transferred to another building that is underheated. (The university expects to save $2 million a year in energy costs through this project.)[34]

Among stand-alone projects, anyone who wants a glimpse at architecture of the future should consider the Oregon Sustainability Center, a high-rise building set to begin construction near Portland State University in 2012. The $65-million, 130,000-square-foot building, supported through money from the Oregon University System, will be the largest to strive for the Living Building Challenge, the most stringent green building program in the world.[35] (Most Living Buildings are approximately the same size as one constructed by Washington University in St. Louis in 2009 that was 2,900 square feet.) The Living Building Challenge was set up by the Cascadia Region Green Building Council, a Northwest chapter of the U.S. Green Building Council. It is an all-or-nothing, sixteen-point rating system that requires a building to produce its own energy, capture its potable water, and clean or recycle its wastewater. A building must use sustainable, local construction materials and avoid a "red list" of prohibited materials that includes substances as common as polyvinyl chloride (PVC). Moreover, the building must be beautiful, a pleasure to work in, and an educational example.[36]

Needless to say, these are very difficult standards to meet, but a financial study by the Cascadia Region Green Building Council indicates that Living Buildings could be erected in most regions of the country if building owners could tolerate higher up-front costs and longer payback periods. The study compared various hypothetical Living Building types in four cities with different climates: Portland, Atlanta, Phoenix, and Boston. Construction premiums and payback periods varied according to building type, local climate, and local energy costs. For example, a university classroom building in Portland would cost $339 per square foot, with a 4–9 percent cost premium and a payback period of two to seven years. The same building in Boston could be built for $395 per square foot, a 16–21 percent premium, and a payback within six to eleven years. Tables turn with, say, a low-rise office building; in Boston, the square-foot cost is less and payback is quicker than in Portland.[37]

Figure 13.4 The Oregon Sustainability Center at Portland State University will strive for certification under the Living Building Challenge. To get that certification, the building will have to produce more energy and capture more water than occupants use, along with achieving other difficult environmental standards. Courtesy of GBD Architects and SERA Architects.

Jay Kenton, the vice chancellor for finance and administration in the Oregon University System, describes the building as "a branding effort" to show that Oregon is a sustainability leader. The motivations here are as much financial as environmental. Oregon has one of the highest unemployment rates in the country, and the university system, state representatives, and Portland city officials hope that the Oregon Sustainability Center will inject life into the state's green building industry, attracting researchers and companies to work with Oregon firms who achieved the project.[38] Wim Wiewel, president of Portland State University, adds, "The Center is a prototype building, so its costs are higher than standard construction, but public and philanthropic resources will be used to bring rental costs to a market level. This subsidy makes sense because the building is seen as an engine of economic development for the sustainability industry in the region. It is a great example of university-city partnership for regional development."[39]

Now architects just need to complete the project. One of the main challenges is generating enough solar power in Portland's climate, which is overcast half of the year. An early design of the building, as a high-rise, did not have a lot of horizontal roof space to work with. A huge solar array sheltered the top of the building, and solar panels were placed on solar shades over windows and balconies below.[40] The university agreed to reduce the size of the building by some 40,000 square feet to ensure that it is properly scaled to meet its net-zero energy goals. "That was exciting," says Clark Brockman, an architect on the project. "Architects don't usually get told to make the building smaller, at least for that reason. Usually we get told to make it smaller for the budget." Portland is at least a more forgiving climate for planning a building that plans to harness the environment. "We used to do this a lot more 150 years ago—buildings were a lot more place-based, tied to where the winds blew and where the sun came from," Brockman says.[41]

With the building scaled down, a greater proportion of the energy use will come from the building's occupants. Nonprofit organizations, for-profit companies, and researchers from Portland State University, all focused on sustainability issues, are the prospective tenants, and the architects have been asking them through online surveys which comforts and conveniences they would be willing to give up to save energy and resources. The tenants have agreed to a number of unconventional practices: to clean the building during the day instead of at night to minimize the time the lights would be on; only one sink with hot water in the bathrooms to minimize hot water heating; to use laptops and remote servers instead of inefficient tower computers. The building may be metered for energy and water use down to the office, even down to the person; the organizations would have to commit to keeping their consumption below a set level.[42] In a building financed by tenant rents, imposing limitations on the occupants concerns Jay Kenton, the vice chancellor for finance and administration in the Oregon University System, who is in charge of the project. "I don't think most people think of sustainability in that way—gee, this really means that we are going to have to do with less," he says.[43]

This may be one of the greatest lessons of all about green building. Our climates, geographies, and living arrangements have limitations; and while solar panels, gray-water systems, and geothermal systems might create more sustainable structures, the solution to the problem of unsustainable building lies with people in buildings, not entirely in the buildings themselves. Many sustainability advocates, journalists, energy experts, and economists have noted that changes in behavior and attitude about consumption of energy and resources could offer the most dramatic and immediate impacts. Addressing that problem on a campus may be even bigger than any single building project, but it is a challenge that administrators will have to face to an extent that they so far have not. Building trends at colleges and universities in the past couple of decades have indulged students, star faculty members, and administrators themselves. University leaders who want to build green may have to start difficult conversations about limitations. We will not only have to build the way we used to build but also live in those green buildings the way we used to live.

CHAPTER FOURTEEN

Sustainable Campus Housing: Building a Better Place

Norbert W. Dunkel and Lynne Deninger

There are approximately 2.6 million students living in college and university campus housing in the United States. Comprehensive housing operations now include business services, facilities management, residence life and education, human resources, housekeeping, research, information technology, academic initiatives, and marketing and public relations. Campuses provide housing for undergraduate students in residence halls, graduate and professional students in apartments, Greek chapter housing, faculty housing, and retirement housing for alumni. Campus housing configurations have evolved from two students in a room sharing a communal bathroom down the hall, to suites where two rooms share a full bathroom, to super suites in which four double occupancy rooms share two full baths, to apartments where four single bedrooms share two full baths along with a kitchen, living room, and dining room. The amenities in campus housing now include standard Internet connectivity, elevators, air conditioning, carpeting, fitness centers, mail service, convenience stores, multipurpose rooms, music rooms, and swimming pools. In fact, many campus housing facilities are built adjoining recreational sporting complexes.

Perhaps one of the largest changes to undergraduate campus housing since 2000 is the resurgence of academic initiatives. These programs, courses, and events have brought the academic elements of an institution closer to its residential component. On many campuses, living and learning communities design a residence hall with a single theme, such as fine arts. In a fine arts living and learning community, for instance, multiple functions and spaces are closely coordinated, including a faculty apartment, a gallery where students can show their work, an acoustically sound music practice facility, a studio in which students can paint, and academic advising and tutorial space, as examples. With this basic understanding of residence life as a background, the following chapter examines the key aspects of sustainable campus housing.

Sustainable Living and Learning Communities: Goals and Benefits

Living and Learning Communities (LLCs) have existed in campus housing for many years. An LLC joins the academic component with the residential component in a focused, intentional living community with a faculty member in residence. These professors may lead travel abroad opportunities, host receptions and speakers, and teach courses in the residence halls. LLCs may have smart, media-level classrooms available for courses associated with that particular LLC. Students living in a specific LLC may have a common academic discipline or simply an interest in the theme of the LLC. The establishment of these academically related residential communities leads to additional benefits for students, including higher grade point averages, increased matriculation to graduation, increased respect and understanding of faculty, and increased psychosocial development skills.

More than six hundred presidents from U.S. institutions of higher education have now signed the American College & University Presidents' Climate Commitment (ACUPCC) with the goal of carbon neutrality. Campus housing operations are positioned very well to lead institutional sustainability initiatives in pursuit of the goal of carbon neutrality. Many campus housing operations incorporate not only the residence life aspects of programming, such as conduct, crisis management, and community building, but also business services, information technology, payroll, personnel, and facilities management, which can incorporate construction/renovation, maintenance services, and housekeeping. Campus housing uses tremendous amounts of utilities to provide the basic services for resident students. The use of electricity, natural gas, and water supports the various lifestyles of students, who in turn produce tons of refuse, sewage, and dining hall leftovers. Campus housing works closely with the institution in the disposal and cost of these byproducts. A number of campus housing operations have developed an effective approach to reducing or eliminating these byproducts by establishing Sustainable Living and Learning Communities (SLLCs).

In 2007, Maruja Torres-Antonini and Norbert Dunkel reported finding eighty-seven campus housing initiatives in U.S. colleges and universities "aimed at either sustainable building, sustainability education, or both."[1] These institutions have made a commitment in campus housing to advance their sustainability efforts through one of three different initiatives: (1) green campus housing, (2) sustainability-themed LLCs, and

Three Emerging Trends

- The development of sustainable living and learning communities
- The re-prioritization of campus housing maintenance and renovation funds toward inclusion of sustainable elements
- The creation of a position in campus housing dedicated to advancing sustainability

(3) campus housing sustainability "hubs." Fifty-nine of these institutions offered campus housing "purposefully designed and operated to meet, as well as model, optimal energy efficiency and overall environmental performance. These residences are green buildings, defined as those that maximize energy, water, and materials use and that minimize and ultimately eliminate negative impacts on human health and the environment throughout their life cycle."[2] Seventeen additional institutions indicated that they possessed a sustainable residential learning community or LLC. They made a commitment to educate the students in their LLC on various aspects of sustainability. These seventeen LLCs may invite a speaker in to share thoughts on sustainability or may require the hall staff to sponsor a sustainability-related program or to take the students to an off-campus recycling site to better understand that aspect of recycling.

Eleven institutions had campus programs that included both green building advancements and a comprehensive student experience. These LLCs included specifically designed physical environments that supported the sustainability education program such as comprehensive recycling programs with containers in each of their rooms, separate trash chutes for recycled materials, and students identified as floor recycling coordinators. The residence might utilize Energy Star appliances, low-flow showerheads, infrared sinks in bathrooms, motion detectors on common area and bathroom lighting, and water-free urinals. The educational program will often include one or more faculty members teaching courses on sustainable topics, students working on internships with local companies to understand sustainability in the private sector, and opportunities for trips abroad to be exposed to the global push for sustainability. When designed effectively, these programs challenge students to leave their residential experience with a new view of their own lifestyle that fully embraces sustainability in both personal life and professional work.

Students increasingly arrive on campus and move into residence halls with the full expectation that they will be met by programs, amenities,

Projects That Exemplify Best Practices

- Emory University's Piedmont Project
- Cornell University's Ecology House
- Duke University's Smart House

and services that support sustainable practices. Many students have been raised or educated with an ecological literacy and with basic recycling, re-use, and conservation efforts as part of their everyday lives. As entering first-year students, these are the ones who held leadership positions in sustainability clubs and organizations in high school and who led community-wide efforts toward sustainable goals. They are seeking a higher level of involvement, commitment, and understanding on their college or university campus and are willing to dedicate time and focus their energies to educate and motivate other students, even upper-class students, toward these same goals.

Environmental educator David Orr encourages us to "capitalize on the educational power of buildings by using the collegiate campus as a 'tangible model' of sustainable practice. The campus housing sustainability hub (LLC) has the potential to be this model. It combines the emergent pedagogies of active and collaborative learning in higher education and the current redirection of interest in academia toward eco-centric concerns." Orr remarks that "the question is not whether colleges and universities could help catalyze the transition to a sustainable society, but whether they have the vision and the courage to do so."[3]

Sustainable Campus Housing Programs in Partnership with Faculty and the Curriculum

Assuming that universities and colleges should be teaching their students about sustainability, what is the best approach? Some activist groups, such as Greenpeace, who feel that urgent action is necessary, would likely support a contemporary critical-pedagogy approach. In his book *Critical Pedagogy,* Joe L. Kincheloe describes critical pedagogy's central dynamics: "Advocates of critical pedagogy are aware that every minute of every hour that teachers teach, they are faced with complex decisions concerning justice, democracy, and competing ethical claims. While they have to make individual determinations of what to do in these particular circumstances, they must concurrently deal with . . . the surrounding institutional morality."[4]

Critical pedagogies rest on the idea of teaching for change, which is central to the revolutionary change required to transform our collective sustainable future. However, because this relatively "edgy" posture may or may not support the long-term mission, outlook, and values of many higher education institutions, one could say that it is not the most reliable or sustainable approach for change. Experience has shown that the most sustainable or longstanding emerging approaches are those that are derived from student initiatives and that build on the institution's support and mission. As noted by the University of Plymouth's Centre for Sustainable Futures, education must focus on critical, systematic thinking that reconnects students with their environment through an interconnectivity that is only achieved through experiential learning.[5] Experiential learning is action-oriented, includes open dialogue, supports research, and allows for individual and group reflection. Given this, the residence hall environment is the ideal collaborative learning environment for supporting and shaping a collective sustainable future.

In LLCs focused on sustainability, empowerment of the individual student learner is paramount. Several institutions have chosen to ask their students to support their institutional mission in the area of sustainability by signing a pledge or commitment similar to the following, from Emory University:

> While the University has committed itself to achieving overall sustainability, it will take the active participation of the entire Emory community to pitch in, turn off, conserve, and re-evaluate daily habits for Emory to realize its sustainability vision.
>
> We invite you to pledge your efforts to address energy, sustainable food, water conservation, green space, commuting, recycling, and other sustainability issues when at Emory and at home.[6]

At Emory, this approach has met with great success. Students who make such a commitment or accept a sustainability challenge early in their academic careers often become more connected to their residential

A Decision One Would Make Differently with Hindsight

- Include larger-scale sustainable initiatives in residence hall construction projects. Be bold, be a leader, and do not wait for all the data and information to come in to fully support adding an element or initiative before taking action.

Single Most Important Piece of Advice for a New Professional

- Do not wait for the campus president to tell you to move forward on sustainable practices. You are in the best position to move your housing operation forward and lead the institution.

neighborhoods, their schools, and their larger communities through service learning. For example, Emory's Piedmont Project was developed by a group of faculty to address urgent societal issues and environmental and sustainability concerns.[7] Each summer the project draws together groups of approximately twenty faculty members to learn about environmental issues and sustainability and to develop new courses and course materials. The original project has been expanded to include graduate students, and now all second-year graduate students engage in interdisciplinary discussion about integrating environmental and sustainability issues into their teaching and research. A related initiative, "Emory as Place," is described as "committed to living out our commitments to sustainability":

> Our service learning activities begin with campus-based experiences, hikes and clean-ups, planting programs, and removing invasive species. Often, we do these in conjunction with the Friends of Emory Forest. But our larger intellectual and moral goals draw us to service learning partnerships beyond our campus, where the water and energy that leaves our place intersects with the real lives of other communities' down-stream or down-grid. Those community partnerships remain a key aspect of our service learning commitment, to join with others to make Atlanta and the larger bio-region a more sustainable and thriving place.[8]

Service learning and its research opportunities are also excellent opportunities to set students on a sustainable path. Service learning links curricular objectives with meaningful community service experiences. Almost all colleges and universities now embrace this methodology because it not only supports interdisciplinary collaboration but also challenges students to gain a greater understanding of practical, real-world problems and to use newly acquired skills in addressing them.

Evergreen College in Olympia, Washington, also effectively integrates academics with social expectations. From their first day on campus, Evergreen students are placed in a setting that encourages collaboration in all subject areas. Students in all Evergreen programs are taught to apply

A Sustainability Myth

- Investing in sustainability costs too much.

abstract theories in solving real-world problems; in turn, this approach develops personal independence and critical thinking, essential attributes for future generations growing up in a global economy. The Center for Community-Based Learning and Action is Evergreen's newest public service center. The center traces its roots to the core of Evergreen's distinctive mission by encouraging personal engagement and collaborative work. A tremendous example of community engagement is exemplified by the Gateways Program founded by faculty emerita Carol Minugh, "which has turned college into a reality for incarcerated youth for the past fourteen years. The full-time academic program brings Evergreen students and faculty to detention facilities as mentors, tutors, and instructors to the young men held there."[9] Speaking of the program, Evergreen faculty member and forest ecologist Nalini Nadkarni commented, "It seems like prisons and Evergreen are strange bedfellows, but actually we're not. . . . If we're heading students toward careers that involve service and helping improve the world, then corrections is an extremely logical place to be."[10]

Ms. Nadkarni, collaborating with the Washington State Department of Corrections' Dan Pacholke since 2004, has helped implement sustainable measures at Cedar Creek Corrections Center. The first challenge they tackled together was to cut the facility's water consumption:

> They built a recycling shed out of recycled wood. They didn't put in a fancy $500,000 composter. They put in a low-tech worm operation that does the job a whole lot better. The results were dramatic. They dropped per capita water uses from 132 to 100 gallons a day and saved 250,000 gallons in the hot season alone. And because they were pumping less water to the sewage plant and had taken the seemingly trivial step of scraping dinner plates for the compost operation, the sewage treatment plant was running well below capacity and the need for an expensive fix disappeared. "It was all done on a handshake and a shoestring," Pacholke says.
>
> Now Evergreen and DOC have entered a formal agreement to bring staff and students together under Nadkarni's and Pacholke's leadership to keep the work progressing at Cedar Creek and build out the sustainability capacities of three additional prisons: Stafford Creek, McNeil Island, and the Washington Corrections Center for Women in Gig Harbor.[11]

Gateways has evolved into a remarkable example of students, faculty, and staff members engaging in a service-oriented project for the betterment of a historically disadvantaged population.

Sustainable pedagogy can directly facilitate the development of new methodologies in collaborative settings as evidenced by programs at Cornell University and Duke University: Cornell's Ecology House is devoted to its residents' shared commitment to environmental stewardship. Students follow ecologically sound practices in the house and regularly organize educational outreach programs and activities for the entire Cornell campus.[12] At Duke, the Smart Home was built.

> Designed by Duke University students through a strategic partnership with The Home Depot, the 6,000-square-foot home features a variety of eco-friendly and high-tech elements and will house ten students. . . . Duke students and their advisors designed The Home Depot Smart Home to be adaptable, innovative, environmentally sustainable, and technologically integrated. . . . Once they move in, the 10 residents will automatically become ambassadors of sustainable lifestyles, conducting tours and answering questions about energy-efficient, environmentally responsible living in a stylish, high-technology setting. . . . The residence hall and research laboratory is the centerpiece of a much larger program in which more than 100 students are conducting research on smart living.[13]

Net-Zero Living: Our Path to the Future

Concern about the impact of carbon dioxide on the environment has caused increased concern and discussion on campuses over the past decade or more. According to Laurence O'Sullivan, "Most of the observed increase in globally averaged temperatures since the mid-20th century is very likely due to the observed increase in anthropogenic (human) greenhouse gas concentrations. Anthropogenic warming and sea level rise would continue for centuries due to the timescales associated with climate processes and feedbacks, even if greenhouse gas concentrations were to be stabilized."[14] Since the climate-change discussions have become an international phenomenon, many U.S. campuses are actively trying to reduce their greenhouse gas emissions. Faculty and staff can make a difference by changing daily habits and behaviors.

In university and college architecture, these concerns have taken the form of Ed Mazria's 2030 Challenge, which targets carbon reduction goals in five-year increments culminating in carbon-neutral buildings by the year 2030 and, more recently, with "net-zero" living within a net-zero-energy building, that is, a building that produces as much energy as it consumes. Campus residences can be designed for net-zero energy consumption with little to no sacrifice, as evidenced in the examples discussed here.

Net-zero living starts with net-zero construction, which embraces passive energy features that reduce energy demands. Effective passive energy features that are well suited for campus residences include the following elements:[15]

—*Passive solar heating* uses sunlight to heat objects within an occupied space. South-facing windows are the most effective for maximizing heat gain. Once heat has entered the building, various techniques are used to maintain and distribute it. Concrete or stone flooring or walls with heavy mass can be used to absorb and distribute heat. Up to 25 percent of a building's heating requirements can be gained with passive solar techniques.

—*Passive cooling* reduces or removes heat that accumulates in buildings through four main strategies: (1) natural ventilation, (2) evaporative cooling, (3) high thermal mass, and (4) high thermal mass with night ventilation. Cooling can also be accomplished with a high degree of insulation, high-performance glazing that blocks heat gain, building orientation, and exterior shading. To determine which passive cooling strategies are appropriate for a particular building site, a bioclimatic chart should be reviewed.

—*Daylighting* is controlled admittance of natural light into a space through windows. It can reduce or eliminate the need for electric lighting and create a more stimulating environment for building occupants. Up to one-third of total building energy costs are related to lighting, and significant saving can be identified with daylighting strategies. Daylighting has the potential to significantly improve life-cycle cost, increase user productivity, reduce emissions, and reduce operating costs up to 30 percent.

—*Superinsulated exterior envelopes,* especially spray-on options, are most effective in reducing energy loss due to air leakage as well as controlling moisture and increasing human comfort while minimizing the need to heat and cool the space. This approach, combined with other low-energy building technologies such as high-efficiency heating, ventilation, and air-conditioning equipment; natural ventilation; and evaporative cooling are the foundation of long-term net-zero buildings.

All buildings consume energy, however, so for a building to be net zero, it must either generate energy itself or contribute to the generation of energy elsewhere in order to offset its energy consumption. Specific opportunities to achieve this goal include one or more of the following components:

—*Solar photovoltaic (PV) panels* can convert the sun's rays to electricity and heat a building and its hot water supply.

—*Micro-hydro or micro-wind energy* can, in the right location, provide from five to ten KW.

—*Stand-alone power systems (SAPS)* can provide off-grid electricity generation that may include photovoltaics, turbines, batteries, and a back-up generator.

—*Grid-connected systems* are electricity generation systems that may include solar PV, micro-wind turbines, or micro-hydro, connected to a local electricity network providing the opportunity for electricity exportation and reuse.

—*Geothermal alternatives* take advantage of the fact that ten feet of the earth's surface maintains a nearly constant temperature between 50° and 60° F, which, combined with geothermal heat pump devices, can leverage the earth as a source of heat for both heating and cooling.

Beyond the energy savings inherent in their architectural design, sustainable residence halls produce other benefits. By facilitating LLCs, these buildings and the experience of living in them can profoundly affect the mindset of residents, educating them about sustainability thinking, increasing their compliance with energy-saving mechanisms, and imprinting them with a concern for the environment that can last a lifetime—a particularly desirable quality to cultivate in tomorrow's policy and decision makers.[16]

By increasing students' awareness of their own consumption of resources and encouraging them to conserve energy and water, sustainable living-learning communities can go a good distance toward achieving net-zero campus living. College and university residences, like most homes, devote the largest share of their energy consumption to heating, cooling, and hot water, as shown in figure 14.1.[17]

Student residents can help reduce heating and cooling energy consumption by wearing more clothing and turning the heat down slightly. Turning the thermostat down by one degree can reduce carbon emissions and cut fuel bills by up to 10 percent.[18] For example, students at the University of Tennessee–Knoxville are supporting efforts to reduce energy consumption by 10 percent in the 2009–10 fiscal year to save

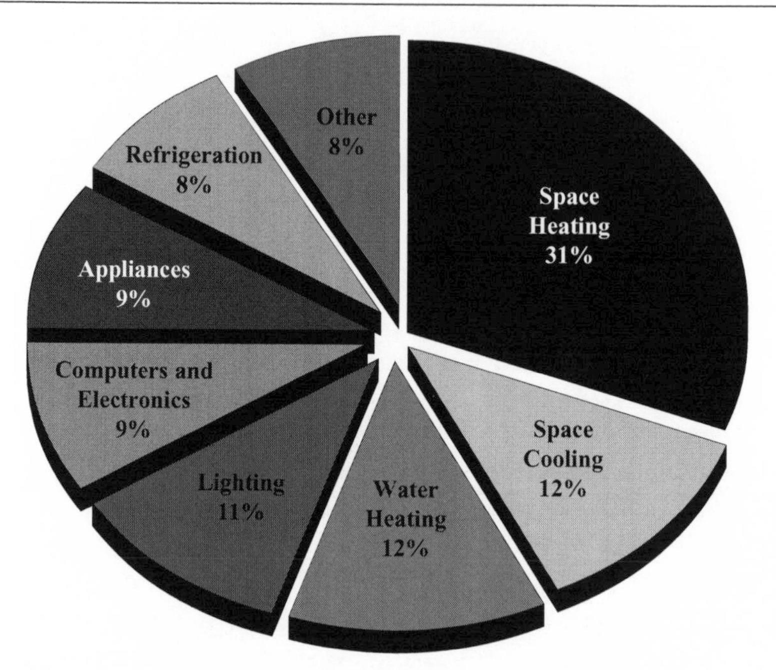

Figure 14.1 Heating accounts for the biggest chunk of a typical residence's utility bill; university housing is no different, only at a larger scale.

approximately $2 million. This will be accomplished by embracing a new energy conservation policy that changes temperatures in the majority of buildings on the university's campuses. During normal occupied hours, the target indoor air temperature will be 68 degrees Fahrenheit for heating and 76 degrees for cooling. The campus is encouraging dressing in layers and embracing solar heating.[19] Beyond heating and cooling, the biggest impact students can have toward net-zero living are in the areas of hot water and electricity use. As noted by the American Council for an Energy Efficient Economy, six activities use the most hot water (see table 14.1).[20]

As discussed earlier, an increasing number of American higher education institutions are asking their students to sign pledges confirming that they are willing to make these types of personal changes, semester to semester and year to year. The prospect of direct, personal accountability increases the opportunities for success. A prominent early example of this kind of success has been recorded at Plymouth State University in New Hampshire:

Table 14.1. Hot Water Use in a Typical Home

Activity	Gallons per Use
Clothes washing	32
Showering	20
Bathing	20
Automatic dishwashing	12
Preparing food	5
Hand dishwashing	4

Reducing water use from these typical levels is one of the ways an average family can reduce utility bills for water heating.

Plymouth State University students saved nearly 12,000 kilowatts of electricity in a very successful energy-saving competition. Danielle Dustin, Samuel Read Hall residence director who organized the "Do It in the Dark" contest, reports students saved 11,764 kilowatt hours of energy during the one-month competition that ran during the month of November. Dustin reports the combined energy savings from all residence halls are the equivalent of 196,067 60-watt light bulbs burning for sixteen hours each.[21]

Campus transportation options provide a final opportunity to reduce students' carbon footprint both while they live on campus and after they leave in terms of their lifestyle decisions. Campus-wide transportation such as shuttle buses, especially if they are powered by biodiesel fuel, can achieve dramatic carbon footprint reductions, and bike rental and bike sharing programs on campus can have a noticeable impact. Furthermore, there is no mode of transportation more user-friendly or sustainable than walking. Having a safe, well-lit, clearly marked campus encourages people to forego cars and walk to their destinations. Although removing automobiles from campus altogether may be the most environmentally sustainable solution, it is still unfeasible in most situations; however, promoting alternative-energy vehicles, carpooling, and ride sharing are effective methods to reduce carbon footprint while networking the extended campus community. Even with these creative models, net-zero living remains an ideal that requires not only an authentically built environment but also a durable, continuing sense of engagement from those within it.

Conclusion: Campus Housing as a Strategic Resource

College and university campus housing operations are among the best-positioned areas operationally and strategically to advance sustainability initiatives and to reach net-zero living. Operationally, its administrators increasingly seek systems and processes in their negotiations that include environmental, economic, and socially sustainable practices within their missions. Many housing functions, such as heating and cooling systems, refuse disposal, wastewater use, electrical and water provisions, and fire suppression systems, carry environmental implications; and skillful housing directors can derive extensive financial benefits from applying sustainable practices to these functions. Whether the approach is student-driven recycling, collection and reuse of gray water, solar pre-heated hot water, or energy-efficient light fixtures, to name only a few, officers may choose to retrofit older facilities with these technologies or incorporate them into new construction.

College and university campus housing officers are also now focusing to a much greater extent on integrating economic sustainability into their practices. Research studies, such as the 2009 Belch, Wilson, and Dunkel study of cultures of success have been commissioned to study the recruitment and retention of professional live-in staff, and administrators of housing operations are using this data to analyze the accommodations, meal packages, spouse and partner benefits, and other amenities in order to recruit and retain professionals at all levels for longer periods of time.[22] In a related activity, some housing staffs are reviewing their equipment and practices to reduce injuries and subsequent time away from work as well as to increase efficiency and productivity.

Few college and university agencies have the scope and diversity that campus housing operations possesses. It is strongly linked to the educational mission of the overall institution, intimately connected to the health of the enrollment management function of the institution, and a key driver of the academic and social success of students at any given point. Campus housing and its innovative living-learning communities remains perhaps the most important environment in which students, staff, and faculty learn sustainability thinking. Many of the recent efforts to incorporate sustainable practices and designs into campus housing, taken collectively, have made net-zero living an attainable campus housing goal. Aware of this, college and university housing officers are ready to take the lead.

CHAPTER FIFTEEN

Food for Thought: Building Sustainable Food Systems and Healthy Communities

Howard L. Sacks

Introduction: Food as Interdisciplinary Study

Where does our food come from? For most Americans, the answer is easy—from the grocery store. When people enjoy an abundance of relatively safe and affordable food, they rarely think beyond the supermarket aisle. However, in recent years the source of our food has become a significant public issue. Increasingly, our health-conscious society has become alarmed at high rates of childhood diabetes, heart disease, strokes, and other food-related illnesses, as terms like "*E. coli,*" "mad cow," and "salmonella," virtually unspoken a few years ago, form part of our everyday discourse. In our post–9/11 world, the possibility that our food supply could be vulnerable to terrorist attack now links food awareness directly to global politics; while on a more personal level, escalating food prices caused by rising transportation costs are prompting late-night family budget conversations at many kitchen tables.

College and university communities, and their leaders, are taking notice of this broad concern for food systems and the complex web of food growers, processors, and marketers that accounts for what is served in their dining halls and food courts. Across the curriculum, professors from multiple disciplines and their students are now addressing new and important questions regarding food production and distribution. Vegetarian and vegan groups have become prominent campus populations, as students wrestle with the implications of their food choices for personal nutrition, the environment, and social justice. And of course many students (and their parents) still express food awareness via the familiar, perennial complaints about the quality of their institution's food service. American higher education has become preoccupied with the power of local food systems, rightly or wrongly defining them as sustainable, organic, and holistic. Indeed, *Time* magazine proclaimed on a March 2007 cover, "Forget Organic, Buy Local," clearly an indication of the extent to which local foods have entered the popular imagination.[1] Farmers' markets have become practical and trendy Saturday morning destinations, and more and more students and their families have turned into vegetable gardeners, both for recreation and as a response to a pervasive economic recession.

Interest in local food systems also extends beyond personal consumption. Ohio is one of several states to institute policies to establish a sustainable local food system designed to provide a sound economic base for the state's family farmers and to supply high-quality, affordable food to urban consumers. Knox County, home of Kenyon College, in partnership with its surrounding community, has created an ambitious program of using a local food system as a chief means to preserve the area's rural character in the face of advancing urban sprawl.

Beyond their civic value, initiatives like that at Kenyon College can advance a variety of institutional priorities. As a subject, food affords an exceptional opportunity for interdisciplinary study at the college or university level, fostering a reintegration of the curriculum without the creation of new academic programs. Examining the food system generates opportunities for active learning, original research, and creative work, as student contact with those comprising the food system via fieldwork practicums and internships introduces new experiences of diversity and community engagement that go well beyond traditional service learning. In a broader sense, participating in a local food system fosters a profound connection to place, including enhanced town-gown relationships, that is missing in our highly mobile, individualistic society.

Making the Future Local

At most universities and colleges, efforts to build a sustainable food system begin with purchasing local foods for the dining commons. Yet

Four Emerging Trends

- Academic work that involves students and faculty in the study of food
- Co-curricular activity to broadly engage the campus in food issues
- Systematic collaboration with the surrounding community to establish dependable food sources
- An ongoing institutional commitment to the project

what is meant by local? Defining institutional criteria in this regard is critical to establishing clear expectations for purchasing as well as reasonable benchmarks for assessing program progress and success. Local foods are defined with respect to *geography, social relationships,* and *food quality.* Buying local involves purchasing products from the nearest available sources, but how local one can purchase depends on the location of the campus. Institutions situated in agricultural regions can often obtain many products within a few miles from home, while those in metropolitan areas must search hundreds of miles. Kenyon has adopted a "concentric-zone" model in this regard: its first priority is to support farmers in the county within a radius of approximately twenty-five miles from campus. When products are unavailable from the immediate locale, the college purchases from adjoining counties within an approximate fifty-mile radius. A third zone extends out one hundred miles, reflecting Kenyon's commitment to buying products within the state.

At the same time, a food source may not be local just because it's found nearby. Purchasing food from a national distributor that happens to have an office in one's city is inconsistent with most sustainability goals. Conversely, buying bread from a local bakery even if its flour is not locally sourced may provide a fresh product and contribute to the local economy in positive ways. Building a local food system occurs one farmer at a time, one product at a time, and each new addition must be evaluated for its feasibility and contribution to the college or university's mission. Beyond this, a sustainable food system involves more than simple geography; it requires establishing long-term, dependable relationships between producers and institutional buyers. In a global food system, buyers quickly change corporate food suppliers if a problem

Six Best Practices

Local food service can further institutional mission through the following actions:

- Providing a new, more flexible model for interdisciplinary study
- Expanding approaches to engage diversity
- Enlarging opportunities for original scholarly and creative work
- Fostering meaningful experiential learning
- Contributing to a distinctive institutional brand
- Modeling constructive civic engagement

arises or if they find a better price, but suddenly dropping a local producer who bought feeder calves a year in advance of delivery to provide beef to your institution can spell economic ruin for a family farm. Building a sustainable system thus requires that producers and buyers understand each other's needs and commit to an ongoing relationship that ensures a dependable product at a reasonable price for both parties.

Additionally, buying local improves food quality. Generally, locally sourced food will be fresher and more nutritious than products picked unripe or chemically preserved to survive long-distance transport. Buying local can also afford greater product variety than is available from national distributors. A college's definition of suitable local food may thus include considerations of *how* an item was produced. Are the cattle grass fed on pasture? Were herbicides used on the broccoli? These are certainly reasonable questions to ask, and purchasing directly from local farmers enables food service managers to know much more about how their food was produced. A school is also in a better position to make requests about how the food is raised if it has established a dependable relationship with a local producer.

Concern over quality inevitably raises the issue of local vs. organic. The U.S. Department of Agriculture now oversees a program that certifies food products as organic, ensuring that they are largely free of chemicals. There is much debate in the agricultural community about the relative benefits of organic over nonorganic production, and many higher education consumers equate organic with healthier and higher-quality food. However, embracing organic as part of a commitment to food sustainability involves several considerations. First, organic products will likely cost significantly more than nonorganic alternatives. Second, some local small-scale farmers may operate organically but decide not to become USDA certified because of the cost and bureaucracy involved. Third, a dining hall that goes organic by shipping food

Successful Local Food Systems: Six Ingredients

- A strong partnership with the campus food service
- A product-by-product approach to increasing local food purchases
- A dependable infrastructure across the entire food system
- Adequate training of dining service staff in the use of local foods
- Educating campus consumers on the value of local food
- A financial commitment on the part of your institution

Single Most Important Piece of Advice for a New Professional

- In the broadest sense, participating in a local food system fosters a profound connection to place that is missing in our highly mobile, individualistic society.

across the country from factory farms is hardly sustainable; the large carbon footprint of interstate shipping can easily offset such efforts to improve food sourcing. Indeed, it may be better to purchase nonorganic products from local suppliers.

Taking the Holistic Approach

Building a sustainable food system on campus involves far more than dining halls and meal plans. By definition, any sustainable system will actively engage the entire community in multiple ways. Including food sustainability in all aspects of campus life will attract more constituencies and thus ensure the long-term success of the initiative. At Kenyon College, this comprehensive approach is called Food for Thought.

A key challenge in generating a successful sustainable food system is getting people to *think* about their food—what is in it, how it was produced, where it comes from, why that matters—and there are no more credible places to begin this process than in the classroom and the curriculum. For too long, agriculture has been understood as a subject suitable for study only in agricultural programs at large land-grant universities. Until the last decade, these programs relied exclusively on technological innovations and new economic models to address the challenges facing today's farmers. In contrast, liberal arts environments, dedicated as they are to holistic education, appreciate the inextricable link between healthy agriculture and healthy communities—the necessity of putting "culture" back in "agriculture."

Kenyon's course catalogue begins with a section of "Special Academic Initiatives," and it includes Food for Thought. The entry discusses the initiative's educational and public dimensions. Currently, 10 percent of the college faculty considers food, agriculture, or rural life as a significant ingredient in their coursework, and these offerings are listed by discipline. Students examine rural land use policies in a course on practical issues in ethics (Philosophy), explore the significance of food to Asian cultures in a course called "Rice" (Asian Studies), offer a public exhibit of original work on food and culture in an advanced photography seminar (Art), and prepare an authentic meal using products obtained from local farms in an Italian language course on food and cinema (Modern Foreign Languages). Students may combine three of these courses with a paid summer internship and receive a certificate in ecological agriculture from the Ohio Ecological Food and Farm Association, which certifies Ohio's organic farms.

Interest in food simultaneously extends into extracurricular life and the local community. Home-cooked local-food brunches are a good fundraiser for campus organizations, and the event can include tabletop exhibits exploring family farming. Food sustainability clubs—often partnering with environmental groups—sponsor films, panels, and field trips to area farms or processors. Since academic institutions do not supply their own food, building a successful food system also requires active collaboration with the regional community. In Knox County, and increasingly across Ohio, communities are creating local "food councils," whose membership includes farmers, processors, distributors, cooks, and consumers working together to expand local purchasing and to problem solve during this process. Colleges and universities can establish or nurture a food council by offering expertise, technical resources, and a place to meet, often over dinner in their dining hall. Working closely with Kenyon students and faculty, Knox County's council developed a published guide to local food sources; established farmers' markets in every county village; created films, exhibits, and other presentations to raise consumer consciousness; and assisted in the creation of an economic impact study to guide development opportunities.

All these activities require an institutional commitment to local food as consistent with the college or university's mission. Food can advance a variety of collegiate priorities: providing a new, more flexible model for interdisciplinary study, expanding approaches to engage diversity, enlarging opportunities for original research, developing a more distinctive institutional brand, and modeling constructive civic engagement. Achieving these goals ultimately demands that a food initiative become part of the fabric of the institution's annual operating budget and planning priorities. In Kenyon's case, the college has allotted additional funds to pay for the higher cost of some locally sourced foods, and the new campus dining facility is designed throughout—from the loading dock to the servery—to maximize the use of fresh local food.

Building a Sustainable Food System: Six Necessary Steps

Perhaps surprisingly, building a local food system requires planners to go "back to the future" by reestablishing the regional relationships and networks that were commonplace a generation ago in ways that make sense in the twenty-first century. Reintegrating the many parts that make up a food system is central to long-term success: Farmers, processors, chefs, and consumers must appreciate the unique opportunities and challenges all along the food chain to assure a dependable and high-quality food supply.

1. *Establish a strong partnership with your food service.* Whether a school outsources meal preparation or operates its own dining hall, it is imperative to have the food management team on board. Thanks to the growing interest among higher education institutions, many national food service providers are changing their operations to accommodate local food, but managers dedicated to centralized organization and executive chefs unfamiliar with developing menus featuring fresh ingredients are ill-suited to local food initiatives. Colleges fully committed to sustainable local foods may have to demand new approaches or change the dining service provider.

2. *Start with low-hanging fruit.* Most dining services are organized to purchase food from national vendors, so shifting to a local system presents significant challenges. For example, Kenyon's food service prepares 2,500 meals each day in the course of the academic year. When the food service manager needs forty bushels of tomatoes, she can just bring up a national food distributor on her computer and click on a few boxes to place her order. The shipment arrives just when she needs it, and the billing and paperwork are handled automatically through a central office. Order after order, she can be assured of a dependable supply and a consistent level of quality, and if there is ever a problem, the food distributor carries several million dollars of liability insurance to protect the dining service.

However, what if she wants to buy local tomatoes? In Knox County, she can now go to a local produce auction, about twenty-five miles away. Assuming farmers have brought the product she needs, she can spend the morning waiting to bid. Of course, she may have to buy several lots with different varieties of tomatoes, only some of which will be suited to the slicing or dicing required for whatever dish is on the menu. It will be her responsibility to load the product into a vehicle she provides and deliver it to the dining hall. She will likely have to pay cash, she will have to do the paperwork later, and, finally there are no locally grown tomatoes in central Ohio after October 1, just four weeks into the semester. It takes time to overcome these challenges, and the process will advance most easily after a few initial successes. Kenyon began by featuring a local-foods dinner one night a month, shaping the menu to maximize locally available products for that particular meal.

3. *Create a dependable infrastructure across the entire food system.* National food systems benefit from a well-established chain that extends from farm to table, but regional communities may lack one or more connecting links. A university may identify a local cattle rancher, but the nearest kill floor and processor may be hundreds of miles away. Alternatively, the food service may demand federally inspected meat, but the nearest processor may have only state certification. One way to overcome many of these difficulties is to identify a local food distributor who is willing to work with the institution creatively and over the long term to source local foods. That means more work for the distributor, but it may be worth it if your institution represents a significant increase in business volume. Whatever the region, much of the work involved will entail building mutually beneficial relationships with people all along the food chain to ensure a dependable flow of goods.

4. *Train and retrain food service staff members in the new system.* Employees used to a cafeteria system with plastic packages of pre-prepared dishes emptied into warming trays will be unprepared for kitchen work that emphasizes preparing dishes from raw ingredients that are served in small batches on demand. For example, Kenyon's dining staff overcooked its first locally produced hams until they were far too dried out because they were used to nationally distributed hams that were injected with water to increase their weight. Begin the transition with staff orientation sessions that convey the broader purposes of the initiative, and encourage staff commitment by using local culinary knowledge.

5. *Educate your students about local foods.* Building a sustainable local food system on campus will have little impact or value if no one knows about it or learns from it. A critical advertisement for the initiative must be the enhanced quality of the food itself, but enduring acceptance of the program will also require public education, a process that begins with consistently labeling local items on the menu.

Consumers today value familiarity and consistency in the food they consume; a McDonald's burger tastes the same no matter where you purchase it, and the fruit at the supermarket is uniform in color and size and blemish free. Students unfamiliar with the variety of foods available and the steps taken to assure product consistency may at first pass over local products in favor of those that are more familiar. For example, a tabletop exhibit—"Why Don't All the Apples Look Alike?"—can begin to change eating habits in positive ways.

Dining service managers must also rethink what is meant by menu variety. Students who eat all their meals in the same location week after week can become bored with dining hall menus, so introducing variety constitutes a significant challenge. Historically, this has been achieved by offering a wide array of menu items at every meal, but thinking about variety seasonally may mean providing a more restricted menu linked to fresh local products that are available.

Education should also extend well beyond the product itself. Kenyon regularly invites a farmer to the dining hall to "demonstrate" a local product and then join the students for dinner. Students connect more directly with the sources of their food, and farmers gain the satisfaction of seeing others enjoy the product of their labor. Alternatively, a cooking class involving a farmer and master chef might feature a local product. The farmer can explain all that went into raising the food, and the chef can explain the product's positive characteristics. The chef would then teach students how to prepare the dish, which would be featured that night in the dining hall. A short documentary film prepared by faculty and students could complement and extend the dining experience to the farm source. Ideally, a local-food initiative will turn the dining hall into a classroom, educating all who eat there about the interconnections among personal health, ecology, economics, and the sociocultural dimensions of food.

6. *Make the financial commitment.* Particularly in economic hard times, one of the first questions chief financial officers will ask is, "How much do local foods cost?" This is a fair question, and food and sustainability planners will need to answer it straightforwardly. A well-designed local food system can certainly reduce some transportation costs, and creating a more direct connection between farm and table can eliminate some of the middle tier in purchasing. Still, the college or university will pay somewhat more for local food—and for good reasons. Local food is generally better quality than products shipped from across the country or processed by national concerns. Meat bought locally will likely be of a higher grade than that obtained from national distributors and prepared to maximize flavor. Typically, local beef will be naturally tenderized by hanging carcasses for several days before butchering, whereas to speed the process, many factory processors simply inject beef with chemical tenderizers.

More generally, food systems become sustainable when farmers can depend on a reasonable price for their product. In a global food system, farmers cannot set their prices and cannot know what they will make until sale time. Part of building trust with local producers involves establishing how much of a product you will need and determining a price up front. Here, as elsewhere, the incessant demand to "supersize"—obtaining the greatest quantity at the lowest price—has contributed mightily to food systems that are unsustainable. Conversely, paying a reasonable price for local food products can be accompanied by identifying strategies to reduce cost. Flash-freezing local fruits and vegetables at peak season can be cheaper than importing frozen foods out of season and extends the seasonal availability of local products.

Another method to contain costs significantly is to reduce food waste. Students often heap piles of everything available on their trays in the search to find an appetizing taste. As a result, much of the food prepared in the dining hall ends up in the trash. Preparing fewer dishes of higher quality can reduce food waste, particularly when the effort is combined with an effective educational campaign about sustainability. A related success has been the decision by many campuses to go trayless in their dining halls, thus reducing the quantities of food taken and the energy and water needed to wash the trays.

In the final analysis, a campus food purchasing system should maximize the benefits inherent in sustainable practices. Buying local can reduce petroleum use, build farm economic viability, preserve green space, and enhance rural communities. Encouraging environmentally conscious production techniques can improve the quality of our soil and water and enhance wildlife diversity. Each food choice we make, as individuals and as institutions, constitutes a civic act that affects us personally and as community members—the goals of Food for Thought.[2]

CHAPTER SIXTEEN

University Athletics and Sustainability: Start on the Field

Dedee DeLongpré Johnston
and Dave Newport

Winning Is the Only Thing

From the *New York Times* to *Business Week,* the news media is running stories about the "greening" of campus buildings, recycling programs, curriculum, and energy supplies.[1] The green movement is changing the hue of campus life everywhere—except perhaps in the most high-profile venue on many campuses: athletics. Indeed, according to the Association for the Advancement of Sustainability in Higher Education (AASHE), principles of sustainability are being applied throughout campus planning and administration, across areas of operation, in investment strategies, throughout research agendas, and in the classroom.[2] Yet research conducted in 2009 by Mark McSherry reveals that "while nearly three out of four athletic departments reported sustainability initiatives are a 'very high' or 'high' priority for their institution as a whole, less than half (44 percent) of respondents said that sustainability was a very high or high priority for the athletic departments themselves."[3]

Hall of Fame professional football coach Vince Lombardi once quipped, "Winning isn't everything, it's the only thing." Indeed, athletic programs are under pressure to win, and perhaps this pressure trumps the ability to focus on greener outcomes and accountability. However, research shows that professional sports teams are outpacing colleges in the race for the green ribbon. So why are collegiate athletic programs trailing, and is this true across all undergraduate athletic divisions? McSherry's research was focused on the 119 Football Bowl Subdivision (FBS) programs, sometimes referred to as National Collegiate Athletic Association (NCAA) Division I-A (Div-I) schools. In 2008, average attendance at Div-I games ranged from 108,571 at the University of Michi-

Three Best Practices

- *Efficient travel scheduling.* Middlebury College puts the men's and women's teams on the same bus for away games against the same opponent instead of scheduling one team at home and one team away. This cuts travel costs in half but calls for tighter facility scheduling because men's and women's games must be played in succession.
- *Integrating the sustainability plan with campus goals.* At the University of Colorado at Boulder, zero waste in all athletics activities and offsetting carbon emissions from athletics facilities is commensurate with overall campus goals of zero waste and carbon neutrality.
- *Integrating student athletes into greening efforts.* Athletes are highly visible and influential role models—for better or for worse. In 2006, Middlebury College's Nordic ski team became carbon neutral by purchasing offsets for all team activities from practice to competition.

Boxes in Chapter 16 without source notes are by Dave Newport with Mark McSherry, president of ProGreenSports.

gan to 10,638 at Kent State University.[4] Attendance at Div-I games alone averages about seven million fans per season. The championship bowl football games reach additional millions of viewers through televised media. Given the size and relative impact of these 119 programs, the survey results are significant.

Assuming that one of the goals of the campus sustainability movement is to stimulate broader societal awareness, how can campuses raise sustainability to a new priority level within their athletic programs and seize the opportunity to reach these extended audiences? This chapter explores the challenges to integrating sustainability into campus athletics, cites examples of success, and gives recommendations for campuses of all sizes that are planning to take sustainability thinking and action to the next level.

The Business Case for Sustainability

To understand where college and university athletics programs stand within the emergence of the green movement, it is important first to understand what is influencing higher education to "go green." Multiple drivers have spurred the recent proliferation of schools jumping on the green bandwagon: leadership commitments from presidents, provosts,

Three Emerging Trends

- *Consistent with campus trends, athletic facilities will get greener.* Recognizing fans' increasing expectations for a green campus, and in light of decreasing green cost premiums and significant operating cost savings, the pitch won't be the only green feature of future athletic facilities.
- *Green athletics programs will become central elements of recruitment efforts.* Increasingly, prospective students are making buying decisions based on their perceptions of campus green. Smart campus recruiters are combining that penchant with campus sports icons because often, those symbols are what prospective students have seen on TV.
- *Green fencelines around athletic facilities will become much larger.* Expect collegiate athletic programs to increasingly be asked to account for and mitigate environmental impacts not just from team travel to and from games—but from fan travel, construction activities, and suppliers' embedded carbon emissions and waste footprints. The University of Florida launched an innovative program to offset fan travel carbon emissions by upgrading the energy profile of local low income homes, creating enormous good will.

and trustees, demand from students, and general societal interest in issues such as climate change and renewable energies. Likewise, emerging research shows that a significant portion of prospective first-year students make their decisions about which university or college to attend based, in part, on their perception of a school's greenness.[5] With groups ranging from the Sustainable Endowments Institute (SEI) to Princeton Review and *Sierra Club Magazine* publishing indicators of sustainability in higher education, green report cards, ratings, and rankings have increased this public interest in sustainability.[6] While the well-recognized SEI Green Report Card, first published in 2007, profiles the three hundred North American institutions with the largest endowments,[7] the collective flurry of ratings and rankings has effectively made sustainability the newest competitive sport in higher education.[8] As one measure, the growth in the campus sustainability movement can be measured by the number of colleges and universities joining AASHE: institutional membership grew from 149 at the start of 2007 to 1000 by April 2010, with at least one campus member in every state.[9]

With this measurable increase in national engagement, why have not the same drivers accomplished tangible changes in more higher education athletic programs? Prior to the FBS survey, McSherry conducted research on the deployment of sustainability across professional sports organizations. A comparison shows a dramatic difference between the rates of adoption by college athletics and the professional ranks. Professional organizations reported a greater commitment to sustainability across a range of issues. For example, 56 percent of professional teams said key decision makers have a "strongly positive" perception of implementing

Writing Their Own Playbook: An interview with Middlebury College Athletic Director Erin Quinn

DAVE NEWPORT

Going green is so much a part of Middlebury College's philosophy that its integration into the college's athletic program seems like an obvious outcome to Athletic Director Erin Quinn. The college's commitment to carbon neutrality by 2015, combined with leadership from faculty and coaches, contributes to the athletic department's drive for sustainable practices.

Like all departments, athletics faces competing interests in every decision they make. But according to Quinn, that does not mean they back away from the challenge. "We fly sometimes. We make snow. Not everything we do is green—nor can we expect it to be. Other factors come into play: safety, marketing, competitiveness, but we do not let the perfect be the enemy of the good."

Why go green?

Consistent with our school's philosophy. We want to be CN by 2015—and we are part of that. It's what Middlebury is all about.

Who is pushing to do this?

I am. And I am getting lots of requests from students, faculty, staff, and alumni. But we did not want outside constituencies pushing us—we want to do it ourselves. So one of our coaches has taken on being the sustainability guru for the athletics dept too. Bill McKibben is a faculty affiliate with our Nordic ski team—so we have a lot of internal capacity. We want to write our own playbook because it gets more people involved.

Where do you feel resistance?

Maybe some individuals on various teams. But these are random instances. Some do not believe in carbon neutrality. Some do not like doing things just for environmental reasons. But, for instance, rescheduling busses so that men's and women's teams share busses to away games saves money but raises ire among some of the teams. While it works for budget savings—they do not want to do it just for carbon. But the cost savings are worth it and nobody argues with it.

Is there an added expense to go green?

If you buy offsets, it will cost you money. You save money refueling vehicles diesel to vegetable oil. So we saved money there. We save money on transportation. We spent five figures analyzing a turf field vs. artificial—so there was a cost there.

What has surprised you the most?

There is not a large body of best practices to draw on. We ask around for ideas about what to do and get few answers. I am surprised at how far in front we are—but I don't think we're doing anything earth shattering. Ideas come slowly. As little as we are doing, it is more than most.

What are you focused on, moving forward?

One of our coaches is the chair of the college's Environmental Council. We also have an environmental liaison on each team. We keep pushing. It is all about integration in Division III athletics—and that's what we are concentrating on.

sustainability, compared to only 30 percent of collegiate athletic departments. Additionally, 47 percent of professional teams are currently measuring or planning to measure their greenhouse gas emissions, while only 9 percent of collegiate athletic departments are doing so or planning to do so.[10]

While the professional sports organizations recognize the business case for sustainability—reduced operating costs, increased fan loyalty, and potential increased revenues from sponsors—many of the college athletic associations have apparently not yet fully embraced the change.

Closing the Gap

Following the release of McSherry's collegiate report, organizations and campus sustainability leaders began to explore reasons that university athletics might be behind both professional athletics and general college and university efforts. McSherry's research, published by AASHE, also got the attention of the media. Publications ranging from *Inside Higher Ed* to *University Business* picked it up and opined that sustainability ought to be among the many interests that athletic programs address.[11] To administrators on Div-I campuses, the gap is rather obvious. Athletic departments on most of these campuses operate on fundamentally different business models than the rest of the institution. While overall institutions are measuring success against metrics such as admissions yields, research funding, and career placement for graduates, Div-I athletic programs are measured against weekly wins and losses, ticket sales, recruitment, and sponsorships. Athletic programs are driven by good public relations. Administratively, most Div-I athletic programs are "auxiliaries"—operational units that have the capacity to derive income and therefore to operate as financially independent entities. Unlike the sometimes painfully slow and measured pace of campus programming, athletic departments operate on short time horizons for decision making and action. During season play, athletic programming is governed by network air time, demands from sponsors, jockeying for tournament and bowl bids, and, in the case of football, the real-time logistics of hosting tens of thousands of fans on campus six or seven times in three months.

Without wins, some athletic programs fail to raise the financial sponsorships and donations that allow them to pay for operations and recruitment. Adding to this, institutional circumstances vary widely among the Div-I schools. While schools toward the bottom of the pack need to fight for dollars and must convince faculty that athletics deserves a role on campus, programs at the top can operate more autonomously without having to justify their value to the campus community. The differences in these metrics and modes of operation are significant when it comes to the way sustainability staff members communicate with athletic program directors and operational staff. Acknowledging these differences is the first step to promoting partnerships based on sympathetic understanding and cooperation. Athletic staff members may have to become more patient as sustainability advocates on campus develop an understanding of the special challenges athletics programs face.

On campuses that already enjoy strong public relations support for sustainability goals and objectives, athletic departments are often drawn in under the broader campus sustainability umbrella. At two institutions examined in detail in this chapter, the University of Florida and the University of Colorado at Boulder, the athletic departments joined with campus sustainability efforts because, finally, they assessed that there was no down side. From a public relations perspective, in fact, there was only an up side.

Single Most Important Piece of Advice for a New Professional

- Put sustainability into your business model, speak it in business terms, and continuously push for more innovative sustainability initiatives.

A Sustainability Myth

"It costs too much."

First, not true. When the University of Colorado at Boulder removed public trash cans and converted concession operations to zero waste, the start-up costs to capitalize compostables, reusables, and other equipment came in around $20,000—but most of that was equipment that will be reused for years.

Second, going green attracts new revenues as green businesses are looking for new markets in which to promote their products and services. Green food giant Whitewave Foods and earth-marketing Toyota both signed on to support Boulder's green sports with significant sponsorships that happened because Boulder went green.

Greening the Gator Nation

The University of Florida Gators won back-to-back championships in 2006 and 2007 in NCAA Men's Division I Basketball and won the men's football BCS National Championship titles in 2007 and 2009. With four national titles in as many years, the prowess of the athletic program is undisputed. Yet during the same period, the university also developed highly successful complementary recycling and climate action programs. In 2006, moving toward its goal of reaching zero waste by 2015, the University of Florida's Office of Sustainability launched the "TailGator Green Team" initiative—an unprecedented game day recycling effort. Student and staff volunteers distributed recycling collection bags and interfaced with tailgating fans across nearly 650 acres. In the first season, the collection effort yielded over 17,000 pounds of recyclable material. The initial game day collection drive was a collaborative project led by the university's Office of Sustainability that drew on resources from the physical plant department, the local chapter of Keep America Beautiful, and the University Athletic Association.

The effort followed the same three shifts for each game. During the first shift, volunteers handed out transparent blue recycling bags and interfaced with fans, while physical plant staff distributed trash boxes and opaque black trash bags. The color difference in the bags provided a visual cue for action. During the second shift, volunteers handed out additional bags, checked existing bags for contamination, removed any non-recyclable materials, and tied off full bags. In the final shift of the

Ten Tips for Greening Athletics Programs

1. Most importantly, athletic department leadership should be educated about sustainability issues and committed to the cause. Executive-level leadership and responsibility for departmental sustainability initiatives will be the greatest factor in success.
2. Form a cross-functional "green" team within the athletic department. Encourage department representation on the campus-wide sustainability team to leverage expertise and to coordinate programs.
3. Develop a strategic sustainability plan for the athletic department with short- and long-term goals, business analysis, and organizational and staff requirements. Clearly define responsibilities and integrate goals into performance metrics.
4. Measure the athletic department's greenhouse gas emissions and other ecological impacts. Set quantitative reduction goals (e.g., GHG emissions, water use, waste, recycling rates) and time lines.
5. Assess fan, employee, and student-athlete interest in environmental issues via surveys and focus groups.
6. Assess new revenue opportunities such as fundraising for sustainability initiatives, corporate sponsorship, and green advertising.
7. Actively engage athletic department employees, student-athletes, teams, and the student body in environmental initiatives.
8. Be authentic and be forthright about your eco-faults.
9. Create active and visible green initiatives that continuously "touch" fans. Big splash announcements without ongoing development and visibility of the green program will be largely ineffective.
10. Differentiate your program. There are still plenty of opportunities to be "the first athletic department that. . . ."

From Julian Dautremont, "Ten Tips for Developing a Sustainability Game Plan for Athletic Departments," AASHE Campus Sustainability Perspectives Blog, 30 July 2009, www.aashe.org/blog/ten-tips-developing-sustainability-game-plan-athletic-departments.

day, one hour after kick-off, volunteers removed contamination from remaining bags and tied them off. Physical plant staff members collected the bags in the very early morning hours following each game. University police also reinforced the message while making their rounds by reminding fans to use the bags appropriately. In subsequent years, the program was cosponsored by the Pepsi Bottling Group, which helped to reward student groups whose members gave their time to the expanding program. The team of volunteers grew to more than a hundred across the three-month season. Sporting their highly recognizable "Green Team" tee shirts, they often earned boisterous appreciation from thousands of fans. Collection efforts improved game after game, yielding a 50 percent increase in collection during the second year. Athletic Association facilities staffers added materials separation to the post-game stadium clean-up contract as well, which added several tons of recyclable material to the final total.

Building on this highly publicized success, in 2007 the Natural Resource Defense Council contacted the university about hosting the first-ever NCAA *carbon neutral* football game. The game, a longstanding in-state rivalry between the University of Florida and Florida State University, was a nationally recognized opportunity to weave tradition and competition into the university's new climate action goals. University of Florida president, Dr. J. Bernard Machen, had signed the American College & University Presidents' Climate Commitment (ACUPCC) in October of 2006, and Florida's governor, Charlie Crist, a Florida State University alumnus, was also taking a leadership role in climate action planning at the time, having hosted a global climate summit in the same year. This high-level leadership helped set the stage for action.

The first step in the effort was measuring the footprint of the game. University officials calculated the carbon emissions that resulted from the operation of the stadium, the estimated travel impact of the 90,000 fans to the game, air travel to the local airport attributable to the game, the emissions associated with the operation of the hotel rooms in which out-of-town fans stayed, and finally the travel emissions generated by the visiting team, the visiting band, and the visiting cheer squad. Next a team of researchers from the university's School of Forest Resources and Conservation calculated the number of acres of young pine plantation forest that would have to be managed over a ten-year span to sequester the estimated carbon emissions from the game. With the calculations in hand, the Natural Resource Defense Council brokered the deal with local landowners to secure the necessary offset credits.[12]

Unbeknownst to university hosts, two fans attending this inaugural carbon-neutral game would set a new pace for carbon-neutral athletics for years to come. A former Florida Gator baseball player and his son were thrilled to learn about the effort to wed environmental action with college athletics. Subsequently, this family developed a proposal that would create, in the following year, an entirely carbon-neutral football season. The Neutral Gator campaign, as it was named, took the initial data generated from the Florida vs. Florida State game and extrapolated it to calculate the footprint of an entire home football season. Through fundraising and fan support, the Neutral Gator campaign generated equivalent emissions reductions by investing in energy efficiency projects in low-income neighborhoods in the community surrounding the university. Campaign leaders also assisted with tree planting projects in the Everglades through a partnership with another prominent Gator alumnus.

The cascading effects of that first carbon-neutral football season flowed throughout the athletic program. The university's vice president for finance and administration, the athletic director, athletic facilities staff members, and officials from the university's foundation came together to learn about the program and to hear the Neutral Gator story from the impassioned former baseball player who made it possible. As a result, key university administrators and trustees were persuaded by the value that this program, and the recycling programs, added to their bottom line. Soon after, athletic department staff members began to take a new leadership role in sustainability efforts. The university's indoor sports venue, the Stephen J. O'Connell Center, integrated recycling and waste reduction strategies into its operations and purchasing policies. The remains of the championship basketball court were spared from the landfill and were instead cut up and redistributed to local craftspeople, who reused the wood for arts and carpentry needs. The generators used to power the university's annual homecoming ceremonies were powered by bio-diesel, and the women's volleyball team hosted its first green-themed game, bringing new fans into the action. Through the support of the Neutral Gator campaign, the university hosted the first-ever carbon-neutral athletic season during the 2009–2010 year—the culmination of two years of carbon action planning. The calculations for the season came into alignment with the university's carbon footprint under the Presidents' Climate Commitment. Instead of calculating fan travel and other Scope 3 emissions, the university measured the operational footprint that resulted from home games as well as the emissions generated by university-sponsored travel to away games. This shifted the focus to strategic changes that could be made in operations, fuel choices, and modes of travel.

All this activity has also attracted some new fans to the Gator Nation. The director of the office of sustainability reports that some students, particularly graduate students who had not been predisposed to follow athletics, found a new affinity with a program that shared their values. Whereas in the past these students may have watched a few games with their more loyal friends, now they had themselves turned to supporting the new "greener" Gator Nation.[13] Completing the athletic program's bottom-line enhancements, sponsors looking to align themselves with sustainability-oriented organizations began to seek partnerships. For example, a waste management company sponsored all new recycling collection bins for the football stadium in 2009.

Beyond sustainability programming, building and design advances at the university have also earned national notice. The new Heavener Football Complex, completed in 2009, was the first LEED platinum building of any kind constructed in the state of Florida and the first athletic facility with this designation in the United States. The addition houses a trophy room, a recruitment center, and training facilities. The university's commitment to build all new construction to a minimum of LEED gold standards opened the way for this effort. Meeting these new standards was not cheap; however, the features will help pay for themselves through reduced long-range operational costs.[14] The success of the project is also attributed to the university's LEED project manager. She considered all primary options to earn LEED points on the project, and without her enthusiasm and energy, the project may have gained only gold certification.[15] Efforts to take sustainability to the highest levels of achievement must be championed by many individuals and teams across a campus, from the president and trustees outward.

Ralphie's Green Stampede: The University of Colorado Approach

The University of Colorado at Boulder (UCB) added deep green to the football team's silver and gold colors in 2008. Carbon neutrality and the nation's first zero-waste FBS football stadium headlined an otherwise disappointing season, but the university's football stadium, Folsom Field, became the first NCAA Div-1 stadium to eliminate public trash cans and transition to a zero-waste system. At the same time, the stadium and the team attained carbon neutrality for the entire home and away schedule. Thus, UCB became the first carbon-neutral, zero-waste football program in the nation.[16] The concept of zero-waste football events was first piloted at the University of Florida in 2003 and 2004. Drawing from the lessons at Florida and a smaller-scale zero-waste football experience at the 10,000-seat University of California at Davis stadium, zero-waste football operations at the 56,000-seat University of Colorado stadium were first proposed in 2006.[17] Initially, it was contemplated as a literal accomplishment: "zero-waste by 2010." However, it was impossible to foresee a 100 percent zero-waste system, given the amount of uncontrollable wastes that might appear from the visiting teams, shipments of materials to the stadium, and various rogue materials that fans and staff might generate. Accordingly, the vision of zero-waste transitioned into a continuous improvement goal similar to a zero-accident goal on a construction site. With this revisioning of the goal to an aspirational model, the zero-waste effort at UCB was born in 2007.

The UCB effort was a massive scale-up of practices the university had been refining in smaller events for several years, and the scale-up was not without its difficulties. The football stadium, while an appropriate venue to implement zero-waste because it is a closed system with a security parameter, is also managed and operated by a number of different organizational units. Hence, coordination of the eleven different staff units became the central challenge. In addition, for zero-waste to work, all the materials on the retail side of concessions had to be replaced with compostables or recyclables. A challenging price and supply gap between compostables and disposables emerged in 2007, and while much had been achieved in the previous three years, a few difficult-to-eliminate disposable materials still haunt UCB's efforts in this area. Ultimately, UCB's zero-waste system is based on three essential steps. First, working with the concessionaire, Centerplate, Inc., UCB purchases retail food service materials that meet compostability standards.[18] Compared to disposable analogs, compostable products netted a $7,000 difference that included stranded inventory costs in the first year. While it is essential that all materials are converted to compostable alternatives, in reality this is difficult to accomplish. As of 2009, 97 percent of the materials had been converted to compostables.

Second, after removing public trash cans from the stadium, a series of zero-waste stations were installed. These included a can to collect recycling and a can to collect compost, along with appropriate signage and "grade sheets." The stations were hosted by trained volunteers in zero-waste tee shirts and hats who were assigned to talk with fans and to help them understand the system as well as to make the correct disposal choices. UCB recruits and trains about fifty volunteers per game. Third, as bags of compostables and recycling are filled and removed from the stadium, usually starting after halftime, they are inspected by volunteers and staffers, and cross-contamination is removed before being deposited in the appropriate dumpsters. The result is a "clean" compost stream that can be transported to the local facility. Trash contamination is deposited in the recycling vessels and later removed by UCB's recycling facility staff. In that facility, conveyor belts of materials are sorted by paid student staff members, contaminants removed, and the resulting materials marketed.

As a result of these efforts, the first season of UCB's "Ralphie's Green Stampede" zero-waste program spurred a 199 percent increase in recycling at Folsom Field. At most home games, Folsom Field diverted about 80 percent of stadium trash from the landfill into recycling or compost and, in all, recycled 40 tons of material. Green-e Climate Certified Renewable Energy Certificates and local offsets purchased

It's the Right Thing to Do: An interview with Yale University Senior Associate Athletic Director Barbara Chesler

DAVE NEWPORT

There are a hundred good reasons to "go green" in athletics, according to Yale's senior associate athletic director, Barbara Chesler, beginning with its being "the right thing to do." There is a tremendous amount of waste generated in university athletics. The Bulldog Sustainability initiative, as it is called at Yale, provides an opportunity to change that wasteful culture.

Yale's students are heavily involved in the athletic department's sustainability efforts. Student engagement is divided among graduate students, student athletes, and non-athlete undergraduates. In Chesler's opinion, university athletic programs are not leading the way in sustainability efforts across the board because of the cost factor. More than just a simple matter of first-cost expense, the effort requires time and human resources. And unlike Title IX of the Education Amendments of 1972, which ultimately required athletic departments to comply with gender equity standards, sustainability programming is voluntary. "Like Title IX," Chesler says, "sustainability in athletics will take time. It's the right thing to do, but culture change doesn't happen overnight."

Chesler considers sustainability an integral part of her job. Initially, she and the athletic director, Tom Beckett, considered how to launch sustainability efforts within their department. Julie Newman, the university's sustainability director, who was also interested in getting a program off the ground, approached them about working together. Now, several years into the partnership, their cross-departmental team is strong.

The Bulldog Sustainability team has not experienced any pushback from their efforts to date. Budgeting remains the only real hindrance. "With all the money in the world, we could provide all the infrastructure in the world." With more funding, Chesler would ideally hire a sustainability coordinator just for athletics. In the meantime, they are working on marketing the program: they have created a logo and a campaign to raise awareness among athletes, coaches, staff, alumni, and fans.

Retrofitting athletic facilities to increase efficiency in operations is part of the plan, but the cost is generally borne by the university budget at Yale. Unlike on some campuses, where athletic departments are separate legal entities, Yale's athletic department is part of the university. Energy audits led the campus facilities department to make energy efficiency improvements in the recreational sports facilities and the weight room, where 10- to 12-year-old systems had become highly energy inefficient. These changes will save the university more money than they cost to implement.

(*cont'd*)

It's the Right Thing to Do: An interview with Yale University Senior Associate Athletic Director Barbara Chesler

At Yale, the sustainability program is visibly supported by the university president and is implemented by a fully staffed office of sustainability. They have an excellent resource in Jeff Orleans, special advisor to the Council of Ivy Group Presidents, who served as the chief executive officer of the Ivy League athletics conference for twenty-four years.

The sustainability work at Yale athletics is "a long, steady, uphill climb" according to Chesler, who does not see their efforts leveling off in her tenure at the university. They will keep working to change the culture with athletes and with the fan base that attends their contests. She sees tremendous opportunities on the horizon in areas like transportation. With limited parking at their football bowl, the department implemented a park-and-ride system for the Yale/Harvard game last year. Fans park in downtown New Haven and take a university shuttle to the game. Chesler envisions a day when fans come in from outlying posts like Boston on Amtrak or the Metro North line and take shuttles to the bowl—all on discounted tickets.

Time will tell if the Bulldog Sustainability team, formed with a two-year grant, will be able to realize the vision of creating a template for the integration of sustainability into university athletic programs. The pace has initially been slow, but the desire to succeed remains strong.

from the Colorado Carbon Fund offset emissions from stadium energy use and team travel. In 2009, the zero-waste program was expanded to all UCB home athletic events.[19]

Size Can Make a Difference in Sustainability Planning

The results of McSherry's research certainly may have differed had he included Division I-AA, I-AAA, and or Division II and III institutions in the survey. Public relations metrics like fan loyalty and sponsorships do not hold the same weight at these schools. While creating a culture for change at non-FBS athletic programs bears some similarities to Div-I counterparts, the size, scale, and drivers create differences in the models for integration. According to Jeff Orleans, special advisor to the Council of Ivy Group Presidents, educational outcomes are more important drivers within athletic programs at Division I-AA and I-AAA schools.

At Division III Middlebury College, the college's Environmental Council worked with coaches to draft a mission for greening its athletics programs: "In considering Middlebury's athletic and environmental goals, the department of athletics, through its intramural, club, and varsity programs, as well as its physical facilities and interactions with the general public, works to promote a sustainable culture in all sport."[20] Middlebury campus officials report, however, that this campus culture shift has not come without some strife. In 2009 the athletic department, with the help of a campus Environmental Council grant, brought professional athletes to campus for a panel discussion about the conflicts that can arise in making sports more sustainable. The participants on this "Jocks and Treehuggers" panel pointed out that small steps, like recycling, travelling less, and being conscious about the sustainability of the equipment purchased, will become routine over time. These small steps will lead to eventual culture change.

Middlebury's athletic officials have partnered with the school's environmental affairs dean and the campus sustainability committee to work through issues like the sustainability of the gridiron replacement turf, carbon offsets for the women's lacrosse team, sustainable purchasing guidelines for the volleyball team, and the conversion to waste vegetable oil to fuel the ski and crew teams' vehicles. In most cases, the issues were championed by individual athletes or teams. The college's goal to become carbon neutral by 2015 also serves as a driver for change across campus. The intramural sports program has also joined this effort. The director of intramurals reported purchasing replacement equipment for the 2008–9 season from the sustainable Fair Trade Sports Company, and given the large number of participants in Middlebury's intramural program, this purchasing decision had a major impact on this small college's budget.

National Frameworks and Metrics

In April 2008, the NCAA, the organization that guides the governance of the nation's athletic programs, established a "Green Team" composed of national office representatives, athletics administrators from all divisions, and Student-Athlete Advisory Committee members. The organization is now implementing "green" practices at its national office and actively building partnerships with organizations like AASHE while it works on "multipronged strategies to help preserve the environment."[21] The NCAA Green Team is "tasked with guiding the national office staff in supporting the environment, conserving national resources, and developing sustainability efforts. The team also is committed to increasing awareness of environmental issues locally, regionally, and nationally."[22] Beginning with its 2009 convention, the organization has committed to "greening" its annual meetings.[23] What, if anything, could or should the NCAA do to support athletic teams in reaching for sustainability goals? In addition to developing sustainability-related contract specifications for all championship bowl games, the organization can develop and share best practices among programs. Acknowledging that there may be a gap between the desire of athletic staff members to integrate sustainability into operations and their knowledge about how to get started, the NCAA has partnered with AASHE to host an electronic forum for the sharing of best practices.

Additionally, AASHE launched its Sustainability Tracking Assessment and Rating System (STARS) in September 2009.[24] STARS provides a common set of metrics for measuring sustainability in higher education. The framework is similar to the Leadership in Energy and Environmental Design (LEED) metrics developed by the U.S. Green Building Council (USGBC). Shortly after the launch of STARS, the NCAA contacted AASHE to find out how STARS might serve as a leverage point for the greening of athletic programs.[25] Could these metrics, they wondered, assist athletic staff members who wanted to integrate sustainability into their operations? Might athletic associations take a lead in data collection on campuses? Building on the competitive nature of athletics, NCAA representatives are considering whether this reporting system might serve as yet another competitive catalyst going forward.

While some sustainability professionals may be doubtful that athletic departments will take the lead on data collection for one more rating system, many nevertheless agree that a how-to template for athletics is an important starting point. With limited time to research the guiding principles behind action for sustainability, young athletics staffers, in particular, will benefit from a user-friendly blueprint applicable to almost all schools. Some institutions, like Yale University, for example, have capitalized on this kind of co-venture to an even greater degree. In 2008 Yale received a two-year grant to support a partnership between the university's office of sustainability and the athletic program. Objectives of the partnership include "establishing a set of green standards for Yale athletic events that in part reflect best practices researched at other institutions" and "creating a database of best practices collected from a variety of sources so that institutions interested in improving the sustainability of their athletic departments will have an accessible source of the best knowledge and experience."[26] In January 2010, Yale presented its sustainability plan at the annual NCAA Convention.

Final Thoughts on What Works

What lessons can we draw from these examples? In the case of the University of Florida, the director of the office of sustainability notes that strong support from the university president set the tone for success.[27] The university's two aspirational goals—zero waste and carbon neutrality—framed the prioritization of programming. Leadership and passion for change from alumni inspired creativity and reinforced the call to action within the athletic program. The university's vice president for finance and administration is also a member of the University Athletic Association's governing board, providing a bridge for communication across organizations. A clear take-away in this instance is that the athletic association operates as a business with a necessary bottom line. The time scale for action in this environment is much shorter than that on the rest of the campus, yet the sustainability office was willing to take the lead in developing programming, training volunteers, and brokering relationships with stakeholder groups across and beyond the campus. Furthermore, the athletic department was more than willing to play a positive partnership role.

At the University of Colorado–Boulder, a decades-old culture of sustainability championed by students demonstrated to the president and campus leadership team as well as to the athletics department that sustainability pays. Numerous student-led leadership efforts implemented since the founding of a student sustainability center in 1970 have returned great value to UCB, so when a group of students went to the chancellor in 2007 and asked for athletics to join their efforts, it was not long before collaborations began. Sustainability professionals and educators around the world have regularly inquired about or come to observe UCB's zero-waste football games to see firsthand how things are accomplished. UCB's athletics personnel have moved from an initial hesitation about these efforts to embracing them fully. They have witnessed the up side, and are justifiably proud of their success. New major sponsorships have emerged to fund UCB's green efforts, and UCB's athletics director frequently gives presentations and interviews on the how and why of greening athletics.[28] Similarly, the Alumni Association and student government have used these initiatives and institutional momentum to help promote UCB, and these efforts are partly responsible for UCB's having been named the Sierra Club's #1 greenest U.S. campus in 2009.

In our experience, campus athletics personnel involved with greening their programs have noted that there is no appreciable down side to their efforts beyond normal start-up and transition issues and standard management requirements. Furthermore, financial bottom lines will improve over time as cost differentials abate and operational efficiencies are refined. In short, sustainable athletics has the potential to give back far more than is required to implement or maintain it, while at the same time clarifying and strengthening institutional mission.

IV

Beyond the Green Gates: Sustainability and the Institution's External Partners

CHAPTER SEVENTEEN

The Impact of Sustainability on Institutional Quality Assurance and Accreditation

Sandra Elman

> It's not the strongest nor the smartest of the species that survives, but the most responsive to change.
>
> Charles Darwin

Higher Education, Quality Assurance, and Environmental Realities

Environmental and ecological dynamics demand that colleges and universities be active stewards of sustainability. Historically, trustees of institutions of higher learning in the United States have had one overarching mandate: to fulfill their fiduciary responsibility to their respective institutions. With the advent of new environmental realities, boards must now comply with a new mandate: to exercise "sustainable-minded" judgments that inform the formulation of academic and administrative policies that affect the educational programs and operations of institutions. To do otherwise would belie their due diligence. In this writer's view—a view in which she does not claim to represent the views of the Board of Commissioners of the Northwest Commission on Colleges and Universities—understanding the forms and forces of sustainability thinking is not simply one of several options: it is an imperative. Furthermore, it is not simply a political imperative or an ethical imperative; it is a "knowledge-based" imperative.

These are the underlying facts:

1. In 2000 there were ten times as many people on Earth as there were three hundred years ago.
2. The world's population has reached 6.5 billion people. It was not until 1830 that the world's population reached one billion people.

3. There were 263 million people in the United States ten years ago; now the population has surpassed 300 million.

4. In 2003 nearly half of the world's carbon dioxide emissions from cement manufacturing and the combustion of high fossil fuels came from high income countries. The largest emitter is the United States.

For American higher education institutions, sustainability must have a dual focus: sustainability in terms of both its operations and its local and global environments. A college or university must look outward beyond its campus and take into account its numerous constituencies, including those who may never come into contact with the institution but who are affected by its actions and non-actions.[1]

Regional Accreditation as a Catalyst for Sustainability

Regional accreditation in the United States is about sustainable practices, programs, and outcomes. For decades, regional accreditation has functioned as a system of quality assurance aimed at ensuring that an institution is reaching its fullest potential and will continue to do so for the foreseeable future. It is the only nongovernmental, self-regulatory accountability structure that requires institutions to adhere to a set of externally developed criteria that promote good practices. Moreover, it entails a process of continuous improvement. In granting an institution regional accreditation, the accrediting commission is signifying that the institution merits the confidence of the educational community and the public. The primary purpose of the seven regional accrediting commissions in the United States is to protect the public interest.

While accreditation criteria and procedures of regional accrediting commissions differ among the regions, the principles underlying eligibility and levels of expectation are similar in their intent to achieve these goals:

—To foster excellence through the development of criteria and guidelines for assessing educational quality and institutional effectiveness

—To encourage institutional improvement through continuous self-study and evaluation

—To ensure the educational community, the general public, and other organizations that an institution has clearly defined and appropriate educational objectives, has established conditions under which their achievement can reasonably be expected, appears in fact to be substantially accomplishing them, and is so organized, staffed, and supported that it can be expected to continue to do so

—To provide guidance and assistance to established and developing institutions.

The granting or renewing of accreditation signifies that the institution as a whole is substantially achieving its mission and goals and that it meets the commission's expectations for compliance with the region's accreditation criteria. In this sense, the commission's affirmation or reaffirmation of accreditation indicates that the institution is *sustainable.*

The Northwest Commission on Colleges and Universities' new standards for accreditation went into effect in January 2010; implementation of the new standards took place in January 2011. The new accreditation model, which is outcomes-based, requires institutions to take a more systemic and holistic approach to assessing their own performance. The new accreditation model is tripartite in scope. The first part focuses on *purpose and potential* and sets expectations for the articulation of institutional mission and core themes within that mission. It requires identification of indicators of mission fulfillment for each core theme and calls for an evaluation of major institutional functions, resources, and capacity. The second part is concerned with *plans and achievements* and sets expectations for all strategic planning that shapes and reinforces the core themes. Similar to the commission's previous accreditation model, it requires assessment of effectiveness and use of results for improvement. Unlike the pre-2010 model, however, assessments will now need to be conducted with respect to stated core themes, such as sustainability or environmental stewardship, rather than standard major functions. For each core theme, relationships and contributions of resources and capacity are evaluated. For educational programs, goals and their intended outcomes are identified and published.

The third part focuses on *institutional success and vitality.* This section sets expectations for the evaluation of fulfillment of institutional mission as well as expectations for an institution's ability to forecast trends and patterns as it adapts to change. As one illustration of the evolving relationship between sustainability and accreditation, one of the standards within this third part is called "Mission Fulfillment, Sustainability, and Adaptation." It evaluates the college or university's ability to adjust its mission, core themes, and key outcomes to ensure viability, relevance, and sustainability. There can be little doubt that regional accreditation is one of, if not *the* most ideal, mechanism for enabling institutions to be sustainability-conscious entities. Thinking about sustainability requires that institutions see themselves as consumers and beneficiaries of both their immediate physical and broader global environments. This requires a paradigm shift for boards and presidents in governing and administering their institutions. It requires a holistic, systemic approach to decision making that considers the interdependent relationship between the institution and its environment.

Holding Institutions Accountable: A New Standard for Sustainability

One of the many challenges ahead for all quality assurance organizations is to articulate expectations and benchmarks to determine the extent to which institutions are able to demonstrate that they are fulfilling their respective missions, that they are sustainable in an increasingly competitive environment, and that they have the mechanisms and know-how to adapt effectively to external demands and market changes. The Northwest Commission on Colleges and Universities fifth accreditation standard, *Mission Fulfillment, Sustainability and Adaptation,* requires a synthesis of findings leading to an evaluation of the fulfillment of overall institutional mission and goals. These findings are analyzed in the context of internal and external trends and influences that affect the institution's sustainability for the foreseeable future. The notion of sustainability here refers to the institution's ability to sustain itself, financially and otherwise, at a level to adequately offer its educational programs and services and to maintain the caliber of its faculty, administration, and staff going forward. In essence, sustainability refers to the institution's capacity to ensure that its infrastructure is appropriately robust and stable so that students will have every reasonable opportunity to complete their programs of study in accordance with the institution's stated educational requirements and expectations at the time of original enrollment.

This notion of sustainability and its interconnectedness to the fulfillment of an institution's mission is in and of itself a new framework for analysis and evaluation of institutional effectiveness by a regional accrediting commission. However, I suggest that while this holistic culminating standard for accreditation provides institutions with an opportunity to assess their own strengths and vulnerabilities in a fresh way, the new model has also quickly become a *necessary* condition for ensuring the viability of the regional accreditation evaluation process as a quality assurance system for higher education in this country. To be responsible agents of quality assurance and promoters of graduating educated students who will function as enlightened global citizens, regional accrediting commissions and their member institutions together need to view this criterion on sustainability as having a dual emphasis:

A New Standard for Sustainability: Five Expectations

Progressive institutions of higher education can address the broad notion of sustainability by measuring themselves against the following expectations that are embedded but not required by the Northwest Commission on Colleges and Universities' fifth standard for accreditation on sustainability. Within the context of regional accreditation, colleges and universities are able to discern whether they are meeting these expectations:

- First Expectation: Institution demonstrates cognizance of the importance of environmental sustainability.
- Second Expectation: Institution has policies, mechanisms, or good practices that address environmental sustainability.
- Third Expectation: Institution assesses the effectiveness of its environmental sustainability efforts systematically and on a periodic basis as part of its overall institutional planning efforts.
- Fourth Expectation: Institution utilizes the results of its assessment efforts to inform future planning and decision making at the institution with regard to more effectively meeting the spirit and criteria underlying the standard on environmental sustainability.
- Fifth Expectation: Institution adapts environmental sustainability efforts based on results of assessment efforts to ensure heightened mission fulfillment.

environmental sustainability and *fiscal* sustainability. As Charles T. Royer commented in 2000, our "global economy already affects every ecosystem on earth. Just as social scientists are beginning to understand the connection between socio-economic status and health, natural scientists are beginning to understand the connection between human health and the health of the physical systems on which we depend. If our future economic activity—by either its nature or its volume—unduly taxes these natural systems, the health of all populations will be jeopardized."[2]

Demonstrating Good Business Practice: Linking Sustainability and Institutional Planning

The Northwest Commission's expectations with regard to environmental sustainability are embedded in the new criteria for accreditation but are not explicitly clarified. The new standards for accreditation require that institutions engage in planning and implement their plans in accordance with the institution's mission and goals; that these plans be

Myths about Sustainability

It is the responsibility of institutional presidents and accrediting officers, most importantly, to dispel the following myths through discourse, the formulation of effective institutional policies, and the establishment of accreditation criteria that are reasonable and academically sound:

1. *Sustainability means added costs.*
   A number of college presidents and chief financial officers as well as directors of facilities and operations interviewed for this chapter noted that over time, adhering to green practices is not only cost effective but also cost saving.
2. *Sustainability is a fad.*
   Members of boards of trustees and some college presidents maintain that the "sustainability phenomenon" may be likened to a Hollywood-like fad that is occurring in small pockets of business and industry, propelled by creative entrepreneurs who are developing products and wares that are being advertised as keys to saving the planet and enriching lives.
3. *Sustainability is about partisan politics.*
   Some contrarian higher education leaders claim that advancing sustainability is really about pushing a political agenda—usually a liberal political agenda.
4. *Sustainability is an all-or-nothing phenomenon.*
   There is a contingent of higher education leaders who contend that an institution cannot pick and choose what sustainable practices it will follow. Rather, if it is to be "about sustainability," then it has to follow all possible green practices.
5. *Sustainability is tangential to an institution's mission and goals.*
   On the contrary, sustainability is inextricably linked to every higher education institution's mission and goals and is essential to the life of the institution precisely because sustainability and "sustainable-minded" judgments are critical to the viability of each college and university.

holistic, strategic, and grounded in an outcomes-based approach to the assessment of student learning outcomes and institutional effectiveness; and ultimately that institutions determine the extent to which they are sustainable entities. Following this, each university or college will need to make operational, policy, programmatic, and instructional adaptations to ensure its viability in the coming decades. Clearly, higher education institutions cannot operate in a vacuum. Like all institutional facilities, they consume physical resources such as water and electricity and produce waste. They are also consumers of the land and its environs, either as conscientious and laudable caretakers or sometimes as unwittingly wasteful ones. As knowledge-based, teaching-learning institutions, colleges and universities can and should serve as model institutions regarding the consumption of natural resources, energy efficiency, and contributing to the safety, security, and beauty of their campus infrastructure via these institutional planning initiatives:

1. *Incorporate environmental sustainability efforts into the institution's strategic planning initiatives.* This may be manifested in several ways. Physical resource planning and overall facilities planning can be thought of in terms of campus sustainability. In doing so, it can be approached from the perspective of what the Society for College and University Planning refers to as its two core values: "Integrated planning in support of excellence in the academic enterprise" and "innovative, collaborative, and multidisciplinary approaches to planning issues."[3]

2. *Sign the American College & University Presidents' Climate Commitment (ACUPCC).* Presidents and chancellors of colleges and universities who are signatories of this statement are "deeply concerned about the unprecedented scale and speed of global warming and its potential for large-scale, adverse health, social, economic and ecological effects."[4] Numerous presidents in the seven-state Northwest region (Alaska, Idaho, Montana, Nevada, Oregon, Utah, and Washington) have already signed this proclamation. The signatories contend that colleges and universities across the country must exercise leadership in their communities by modeling ways to minimize global warming emissions, and by providing the knowledge and the educated graduates to achieve climate neutrality. They further argue that institutions of higher learning that "exert leadership in addressing climate change will stabilize and reduce their long-term energy costs, attract excellent students and faculty, attract new sources of funding, and increase the support of alumni and local communities."[5]

3. *Participate in the Association for the Advancement of Sustainability in Higher Education's (AASHE) Sustainability Tracking, Assessment & Rating System (STARS).* The STARS system for colleges and universities, as described by its founding manager in chapter 4 of this volume, is a voluntary, self-reporting framework for gauging relative progress toward sustainability. It is designed to

—Provide a guide for advancing sustainability in all sectors of higher education, from education to research to operations and administration

—Enable meaningful comparisons over time and across institutions by establishing a common standard of measurement of sustainability in higher education

—Create incentives for continual improvement toward sustainability

—Facilitate information sharing about higher education sustainability practices and performance

—Build a stronger, more diverse campus sustainability community and promote a comprehensive understanding of sustainability that includes its social, economic and environmental dimensions.[6]

4. *Include sustainability goals and objectives in curricular offerings.* Such offerings can be incorporated into a variety of programs of study in the sciences, social sciences, law, education, humanities, allied health and medicine, vocational/technical programs, and the arts. The biannual ranking of MBA programs of the Aspen Institute seeks to identify innovative programs that "lead the way to integrating issues of social and environmental stewardship into business school curricula and research."[7]

5. *Offer co-curricular activities for students that focus on sustainability thinking and good practices.* These activities could include involvement with the Sierra Club, Habitat for Humanity, local shelters, and local food banks as well as internships with companies that have distinguished themselves by developing a profile in sustainability practices.

6. *Develop an environmentally responsive physical plant operation.* This will include a broad range of activities and initiatives such as recycling of all types, composting food waste, and energy efficient lighting on all campus sites. Residence halls would be fitted with water-saving showers and digital thermostats, and all cleaning chemicals would be "Green Seal" certified. To the extent possible, institutions should work toward mitigating fuel waste and creating a demand for products harvested in environmentally sensitive ways as well as purchasing local products from like-minded vendors. Finally, all new construction, to the degree feasible, should proceed in accordance with the U.S. Green Building Council's LEED silver standard as a minimum. LEED certification, providing independent third-party verification that a building project is environmentally responsible and a healthy place to live and

work, is candidly assessed as a strategic institutional choice in chapter 13 of this study.

7. *Utilize all available campus technology resources in support of sustainability objectives.* One successful approach includes developing a website dedicated to informing students, the college community, and the public about the college's sustainability initiatives along with upcoming "green" events, organizations, products, and courses. An example of this is North Seattle Community College's sustainability website, launched in the summer of 2008 (see http://frontpage.northseattle.edu/sustain).

From Vision to Action: Four Steps for Leadership Teams

Even strong and passionate visions on the part of presidents, provosts, and board chairs are only as good as the necessary operational steps needed to make them realities. So how should leadership teams seeking to fulfill their sustainability responsibilities proceed? The following four action steps provide solid starting points:

1. *Trustees take the lead.* As we approach what Michael Crow, president of Arizona State University, calls a "critical inflection point in the evolution of global society," sustainability issues need to be a priority item on every board agenda.[8] This is why trustee chairs need to demonstrate their leadership and call for a special retreat dedicated to this critical area of concern. The retreat should involve all stakeholders within and external to the institution, including business, legislative, and community leaders. Board members and presidents of peer institutions should also be invited. In this case, competitors may bring useful insight to the topic and provide realistic suggestions to address these issues on a scale that makes sense for the host college or university. The board's leadership then can apply the accreditation criteria that speak to sustainability in shaping a framework for action.

2. *Reposition the institution.* In accordance with its recently adopted accreditation model, institutions accredited by the Northwest Commission on Colleges and Universities are required in their First Year Report to explicate the institution's mission and the core themes embedded in that mission. Presidents often seize this strategic moment to reexamine, through a participatory process, the mission of their institution and its core themes, at the same time providing the community the opportunity to identify sustainability as one of its core themes. Following this, the institution would need to identify specific outcomes to measure the extent to which this core theme and the institution's mission are being fulfilled. Here the accreditation process can provide a pathway to enable institutions

to develop metrics for an outcomes-based, mission-centric evaluative model of institutional performance.

3. *Explore emergent academic disciplines.* The accreditation standards of all seven regional accrediting commissions call for institutions to engage in periodic academic program review. In the Northwest region, institutions are required to evaluate their educational programs and discuss their findings in the Year Five Report of the seven-year accreditation cycle. Chief academic officers, deans, and department chairs can contribute to this campus conversation by developing a range of disciplinary and interdisciplinary programs focusing on sustainability thinking, strategic goals, and operational activities. An example of this model can be found in the launch of the School of Sustainability at Arizona State University. This new enterprise offers undergraduate and graduate programs and "collaborative, transdisciplinary, problem-oriented training that addresses the environmental, economic and social challenges of the twenty-first century."[9]

4. *Plan with an awareness of LEED certification.* Colleges and universities that integrate their planning—specifically, facilities planning, budgeting, and decision-making efforts—ultimately address institutional priorities more effectively and face fewer unanticipated resource shortfalls. Efforts to address sustainability may be enhanced by using LEED criteria. Clearly, many factors and decisions go into creating an environmentally sustainable building and campus. Chief financial officers are often instrumental in guiding presidents and their boards with respect to the cost/benefits of adhering to LEED criteria. Developing a sustainable campus community requires time, focus, and a long-term commitment. Yet incremental efforts can ultimately yield comprehensive results. Higher education institutions can learn a good deal from hospitals, corporations, and planned communities as to how best to design and build a sustainable, environmentally sound campus infrastructure.

Conclusion: The Challenge for Regional Accreditation Commissions

There are no non-governmental entities in this country better positioned and better qualified than the seven regional accrediting commissions to have a substantive and needed impact on the safety, security, and ecological soundness of our nation via sustainability programs, services, resources, and curricula. Together, these regional accrediting commissions directly affect and help shape the policies and practices of more than three thousand institutions of higher education. As the acknowledged gatekeepers for the U.S. Department of Education, these groups are uniquely empowered to affirm practices that encourage effective, sustainable actions while also recommending compliance and areas for improvement where such practices may not be evident.

Ideally, an initiative such as an accreditation standard on sustainability will encourage changes in institutional and individual behavior that can create a more environmentally and economically sustainable future as well as a just society for present and future generations.[10] With member institutions bordering majestic sites such as Mount McKinley, Denali National Park, and the Columbia Gorge, perhaps it is not surprising that the Northwest Commission on Colleges and Universities has been a pacesetter in facilitating the national conversation about sustainability and in enabling higher education institutions to develop and adopt standards for accreditation that facilitate their own viability as well as that of their communities and the nation.

CHAPTER EIGHTEEN

Green Legal: Creating a Culture of Vigilance, Compliance, and Sustainability Thinking

James E. Samels and James Martin

> Now, I truly believe, that we in this generation must come to terms with nature, and I think we're challenged as mankind has never been challenged before to prove our maturity and our mastery, not of nature, but of ourselves.
>
> Rachel Carson, 1963

Rising Pressures on Institutional Leaders

Increasing pressure is now being placed on college and university leaders, including counsel, to meet the legal requirements of long-term sustainability practices and proactively go beyond the *legal floor* to avoid the pitfalls of near-term environmental compliance. In our research for this chapter, we learned from Nilda Mesa, assistant vice president of environmental stewardship at Columbia University, that there are critical distinctions between sustainability thinking and simple compliance,[1] and this chapter addresses key elements of both areas for presidents, provosts, and trustees, in particular.

In this chapter, we focus on the costs, conflicts, and challenges in federal, state, and local regulatory frameworks that higher education institutions face as they implement sustainability thinking and practices. At first glance, some may think it relatively simple to undertake sustainable initiatives on campuses—extra trash receptacles here, recycling bins there—but as many presidents and deans have learned, these goals can be difficult to achieve. For example, in order to comply with state and federal environmental laws, the leadership team must address a complex web of hurdles such as campus power plants, wastewater treatment facilities, asbestos management, research laboratory materials, grounds pollution, athletic facility and medical waste, and hazardous waste disposal, among others.

Noncompliance with state and federal laws can be dangerous to students, faculty, local residents, and the environment as well as extremely costly for an institutional budget, no matter what the size and history of a university or college. In a familiar scenario, colleges may be fined on a daily basis by state and federal environmental agencies; sued by students, faculty, or residents of surrounding communities for damages; or face institutionally incurred damages to buildings and grounds for failing to take proactive measures to dispose properly of hazardous waste. In one instance, during a campus inspection of the University of Hawaii, state and federal environmental authorities reportedly discovered thousands of improperly marked and stored toxic and reactive chemicals beneath a chemistry lecture hall. Had these chemicals been released or spilled, the reaction could have caused a devastating explosion and toxic gas release. This alleged disregard for environmental safety, if not caught, could have caused severe damage to campus infrastructure and student safety. In this case, the university purportedly paid $505,000 in fines and $1.2 million toward environmental pollution prevention projects pursuant to a court-approved settlement.[2]

Presidents and their vice presidents are now more frequently consulting with campus sustainability directors, as well as informed legal counsel, to answer questions such as the following regarding sustainability and compliance concerns:

—Does the college conduct regular environmental audits?

—Is there an annual inspection of campus drinking water for lead and other contaminants?

—Is there a regular inspection of campus HVAC ducts for contaminants?

—How are hazardous wastes from art studios, laboratories, athletic fields, medical laboratories, and laundry facilities handled and eliminated?

—Is the institution a strict user of environmentally friendly products?[3]

In the following section, we offer an overview of federal and state legal systems as they relate to environmental law, compliance, and sustainability objectives. We further examine one of the most common violations administrators must address on their campuses—hazardous waste—including best practices for treatment and disposal options.

The Legal Frameworks Behind Sustainability Planning and Decision Making

As college and university leaders begin to conduct examinations of their institution's environmental practices and safety, it is necessary to understand fully the legal context of requirements of federal, state, and local environmental agencies.

Federal Enforcement and Expectations

Federal enforcement of environmental jurisdiction emanates from the United States Environmental Protection Agency (EPA). The mission of the EPA is to protect U.S. citizens and the environment through a variety of measures, including developing and enforcing governmental regulations, conducting environmental research, and educating individuals and institutions about environmental health and safety. The EPA enforces federal laws, rules, and regulations, including but not limited to:

—*Resource Conservation and Recovery Act (RCRA).* Gives the EPA full control and regulation of hazardous wastes from their inception, to storage, to disposal.

—*Spill Prevention Control and Countermeasures (SPCC).* Requires oil spill prevention, first response preparedness, and response to prevent oil from spilling into navigable waters and shorelines.

—*Clean Water Act (CWA).* Regulates the dumping of wastes and pollutants into water sources. The act made it unlawful to discharge waste into navigable waters without proper permits.

—*Clean Air Act (CAA).* Federal law regulating air emissions from mobile and stationary sources. National Ambient Air Quality Standards have been established by the EPA.

—*Emergency Planning and Community Planning Right-to-Know Act (EPCRA).* Helps protect local communities from chemical hazards. Each state, and regions within each state, has its own emergency plan for chemical disasters.

—*Toxic Substances Control Act (TSCA).* Mandates, in pertinent part, the reporting, proper recordkeeping, and testing of toxic substances as well as provides restrictions related to these substances.[4]

More specifically, the EPA utilizes civil, cleanup, and criminal enforcement. Civil enforcement forces violators to comply with federal laws to eliminate and prevent environmental harm. Cleanup enforcement is used for sites in which hazardous substances have been released or are a threat to be released into the environment, which includes brownfields and land revitalization cleanup projects. Criminal enforcement is used for the most serious environmental violations or instances of gross negligence that involve intentional disregard for the law.[5]

State and Local Enforcement and Expectations

Each state has at least one environmental, conservational, or resource protection agency—with many states reporting multiple entities. These agencies work in conjunction with the EPA to enforce federal as well as local environmental laws. In general, federal regulations establish the minimum national standards. States have the authority to promulgate and enforce stricter standards. In our experience, the most common violations for colleges and universities fall under the above-mentioned federal laws and regulations and include hazardous waste management, oil tank spill prevention, storm water requirements, and underground storage tank provisions. Fines, penalties, and negative publicity constitute what are viewed as final incentives to comply with environmental laws.[6] Common reasons for noncompliance in higher education are:

—Lack of basic knowledge of environmental regulations

—Lack of resources necessary to ensure ongoing compliance

—Lack of financial support from the administration

—Environmental behavior not being given a high priority.[7]

Understanding environmental regulations at both federal and state levels is necessary for higher education institutions to achieve their long-term sustainability goals. In addition, having a basic knowledge of these laws, as well as the most common reasons for noncompliance, can be a motivator for deans, faculty members, and sustainability directors to develop new academic and co-curricular programs to engage greater numbers of students in sustainability awareness and decision making. The University of Massachusetts–Lowell, under the leadership of Chancellor Martin Meehan, a former U.S. congressperson for the state, will open a $70 million Emerging Technology and Innovation Center in 2012. Meehan offers this comment on the importance of linking sustainability with both academic programming and strategic planning: "By design, the University of Massachusetts–Lowell has created an institutional gateway connecting the nanosciences, advanced manufacturing, workplace environmental safety, and sustainable energy resources. In the planning process for our new Emerging Technologies and Innovation Center, a primary goal has been to provide a comprehensive venue for future partnerships in sustainable technologies and the applied sciences."[8]

Dodging the Bullet: When Proactive Planning Is Not Enough

Sometimes, when least expected and with little warning, a university or college may be visited by state or federal agencies to conduct an inspection of its campus facilities. Visits can be prompted by complaints from workers, students, or local community members, or by a specific incident occurring on campus. Whatever the trigger point, the president and leadership team need to be prepared to present auditors with records and permits. The following are useful recommendations to help an institution survive its inspection:

1. Accompany each inspector at all times, following the agreed-upon route, and make the inspector follow the same safety rules as apply to students and staff.
2. Be cooperative, but only supply the requested documents; and if asked a question, be sure employees do not guess or speculate about the answer. Find out the correct answer before responding.
3. Document the visit through notes, photographs, and videos. Afterward, the institution can also request copies of the investigator's notes, photographs, or videos.
4. Be sure to provide the inspector with copies of all requested documents, not originals. Make note to the inspector that these documents are confidential.
5. Should the inspector take a sample of any material, request a split sample in equal volume and weight.[9]

Hazardous Waste: The Four Action Steps

Hazardous waste is one of the most serious legal and environmental challenges for colleges and universities, since the EPA has developed stringent regulations and enforcement of the treatment, storage, and disposal of hazardous waste. The following EPA requirements for dealing with hazardous waste will guide institutions with complying with federal laws:

1. *Treatment.* Treatment of waste means to change its physical or chemical makeup to make the waste less toxic to the environment.
2. *Disposal.* Disposal is to place the waste permanently into land or water, typically in a landfill, waste pile, or injection well.

3. *Storage.* Storage is a temporary solution to the removal of hazardous waste, before it is treated or disposed of. Waste must be stored in proper containers, tanks, or containment buildings.
4. *Permitting, Record Keeping, and Reporting.* Regardless of which method of treatment, disposal, or storage of hazardous waste the institutions uses, the EPA requires the submission of carefully kept records and permits, tracking waste through its lifespan.[10]

Many a campus project has become entangled in regulatory challenges because faculty, staff, and students did not make these special provisions for hazardous waste containment, disposal, storage, and reporting.

Proactive Legal Best Practices for Sustainability Success

In conclusion, we offer these seven best practices for achieving both legal compliance and long-term sustainability goals and objectives.

1. *Do not try to invent the wheel—AASHE already has.* For presidents, provosts, and their legal counsel, in particular, it is useful to remember that an online digest developed by AASHE offers a comprehensive review of campus sustainability efforts in the areas of education and research, campus operations, and administration as well as more than one thousand higher education sustainability stories and new campus resources. In 2009 AASHE launched the Sustainability Tracking, Assessment, & Rating System (STARS), which provides institutions with a standardized and comprehensive way to assess their progress toward sustainability. STARS is described in detail by its founding program manager, Laura Matson, and Judy Walton, in chapter 4 of this volume.

Additionally, the American College and University President's Climate Commitment (ACUPCC), under the leadership of Anthony Cortese, president of Second Nature, has implemented an agenda to accelerate progress towards climate neutrality and sustainability to empower the higher education sector to educate students, create solutions, and provide leadership by example. Signatories of the ACUPCC are required to submit Implementation Profiles, Greenhouse Gas Reports, Climate Action Plans, and Progress Reports in the ACUPCC Reporting System. Public reporting facilitates the ability of institutions to learn from one another. Cortese candidly assesses the overall progress made on these issues as well as the current status of presidential leadership for sustainability action in chapter 2 of this book.

It is also important to remember that colleagues on other campuses are typically willing to share their experience, and this often means going beyond case law and regulatory frameworks to consider past practices, pertinent local customs, and community-based definitions of compliance while acting as good environmental neighbors.

2. *Implement long-term legal environmental policies, procedures, and protocols across the institution with sustainability as a focus.* Importantly, having conducted necessary and appropriate literature searches, legal research, and local 360-degree scanning, ask about new sustainability value propositions, such as projecting costs associated with preventive compliance along with return on earning (ROE), return on environmental projection (REP), and return on sustainability (ROS).

Gradually, our clients' campuses are designating sustainability coordinators—environmental change agents who can influence the creation of a culture of sustainability on campus by encouraging the growth of environmental awareness and literacy within the larger community.

3. *Sensitize faculty, staff, and students to the legal ramifications and liability exposure arising from environmental catastrophes by creating new ways to think about compliance throughout the campus community.* The EPA has a variety of best practices information that includes helpful case studies that address building sustainable programs, energy, sustainable design and building, transportation solutions, waste management, and water management. The EPA New England, for example, has also produced an Environmental Management Guide for Colleges and Universities. This guide was developed in collaboration with faculty and staff members at the University of Massachusetts–Lowell to encourage colleges and universities nationally to design and implement environmental or sustainability programs and a full Environmental Management System (EMS).

4. *Take advantage of the EPA's audit policy.* The best way to protect the institution from environmental liability is to conduct an EPA-approved environmental self-audit. The EPA offers a framework for voluntary disclosure and correction of violations in return for greatly reduced penalties. Incentives are offered to institutions who conduct a self-audit, including protection from an enforcement investigation.

5. *Provide for coordination across departments to ensure that an environmental crisis is properly handled, no matter what operational area it involves.* As an example, some leaders suggest placing compliance issues and sustainability goal achievement on weekly senior staff

meeting agendas and distributing minutes between meetings to encourage ongoing accountability.

6. *Engage in cooperative sharing of legal and sustainability information and resources.* As a primary resource, the National Association of College and University Attorneys (NACUA) offers its members and affiliates access to an extensive menu of legal journal articles, best practice models, and successful sustainability compliance policies that provide newcomers with solutions to unplanned environmental crises. Beyond those of campus legal practitioners, consider using other networks like the National Association of College and University Business Officers (NACUBO), and the Environmental Law Foundation.

7. *Pursue grant funding for sustainable energy and green research initiatives.* This serves the multiple purposes of expanding the institution's overall capacity to "go green," facilitating the discovery of more cost-effective and resource-efficient ways to achieve specific sustainability goals, and enhancing the college or university's ability to achieve compliance where necessary.

Conclusion: Educate, Regulate, Enforce

We advise higher education leaders to wear both cowbells and suspenders when addressing the legal issues related to compliance and sustainability. A simple action plan in this regard includes three rules: educate, regulate, and enforce. The first, *educate,* is to intentionally develop proactive environmental protection and sustainability-driven policies and systematically disseminate this information through multiple publications and campus workshops and training seminars.

The second, *regulate,* requires the institution to promulgate uniform rules in order to create a predictable regulatory environment so that the entire campus community has actual, or at least constructive, notice about the necessary guidelines and expectations in environmental protection and sustainability.

Finally, the third rule is to *enforce.* Presidents, provosts, and boards of trustees need to commit to vigilant enforcement practices because if they do not, federal, state, and local environmental and natural resource law enforcement may cite the institution as a prime example of bad practices and willful noncompliance. As Brian Rogers, chancellor of the University of Alaska–Fairbanks—a state on the frontier of sustainability thinking and practice—cautions, "It is critically important that college and university presidents and chancellors make visible commitments to sustainability initiatives and compliance on their campuses. Our supporters expect it, our students demand it, and our bottom line requires it. We cannot just talk about sustainability; adequate budget support is needed to make a difference. If we fail to make progress, the costs will be reflected in reduced student and philanthropic support and a future far less sustainable than we will need it to be."[11]

V

The Complex Path Ahead

CHAPTER NINETEEN

Conclusion: New Goals and New Challenges for Institutional Leaders

James Martin and James E. Samels

Moving from Operations to Strategic Planning

We opened this book by noting how decisively sustainability goals and objectives—and sustainability thinking overall—had moved from being simply environmental or operational to being strategic on American campuses and how some leadership teams and boards of trustees were still catching up to this trend in part because of a lack of pertinent resources for executive decision making. As one chief executive observed, the problem is that too many in higher education still define sustainability as simply "environmental, scientific, or technological, rather than as an element of the core mission" of universities and colleges.[1] Along with this, presidents and provosts also need to streamline the governance models at their institutions in order to achieve sustainability objectives more quickly and efficiently.

As they relate to programs and practices for a more sustainable future, many colleges and universities claim to be entrepreneurial in purpose and execution, but their reach exceeds their grasp. As sociologist Burton Clark explains, "For a university to be appropriately and productively entrepreneurial, it needs to acquire the right kind of organization, one that allows the institution to be in a state of continuous change. . . . Key features of the new type of organization [are] transforming elements [and] sustaining dynamics."[2] Clark views as transforming elements a diversified funding base and collegial connections between academics and administrators in daily operations. Three key dynamics are mutually supporting interaction, perpetual momentum, and embedded collective volition. Although difficult to accomplish, "these concepts trace the processes by which transformation becomes *sustainable,*" he says, and help universities "to achieve a consequential effectiveness in the twenty-first century."[3]

Experienced presidents and boards realize that it can take years of coordinated effort to put this change model in place amid retirements, career reevaluations, budget uncertainties, and changing student preferences. Clark's description of his first transforming element, a diversified funding base, is prescient. Tony Cortese is quick to agree that the most difficult challenge to overcome at many colleges, even greater than the institutionalization of sustainability thinking, is simply finding adequate funds to stay alive through one more budget year as presidents give in to pressures and spend scarce resources on perennial priorities like deferred maintenance and faculty salaries. In Cortese's view, too many thoughtful, well-documented funding requests are simply travelling down dead-end streets as funders—even some working on campuses—"do not experience higher education as a critical leverage point in the transformation to a healthy, just, secure and sustainable world." Even proactive foundations, for instance, "do not believe higher education institutions will change, or, if they do, that they will change too slowly to be worth future investment. Even more frustratingly, when there are broad-based, collective efforts developed by groups like Second Nature or AASHE, too many foundations still do not believe that they will influence the major elements of the national higher education community."[4] In a glaring instance, Cortese confirmed in a 2010 communication that the American College & University Presidents' Climate Commitment (ACUPCC), one of the most influential sustainability initiatives in higher education, had "been turned down from even being considered for funding by more than twenty foundations over the past two years."[5]

While institutions implement new degree programs in sustainable careers by the hundreds each semester, and as corporations increasingly claim to be the best source for "green jobs" in the region and nation, additional challenges are emerging for presidents and their leadership teams in the extended communities beyond their campuses. Bill McKibben, arguably the nation's most visible environmentalist, expressed some of his most serious frustrations to date in an August 2010 article entitled, "We're Hot as Hell and We're Not Going to Take It Any More," in which he spoke candidly about the realities that recent and future graduates will face as they try to effect lasting progress in our culture related to climate change: "In late July [2010], the U.S. Senate decided to do exactly nothing about climate change. They didn't do less than they could have—they did *nothing,* preserving a two-decade bipartisan record of no action. . . . I wrote the first book for a general audience on global warming back in 1989, and I've spent the subsequent 21 years working on the issue. I'm a mild-mannered guy, a Methodist Sunday School teacher. Not quick to anger. So what I want to say is: this is fucked up. The time has come to get mad, and then to get busy."[6] McKibben goes on to characterize many years of work by groups like the Environmental Defense Fund: "[T]hey wore nice clothes, lobbied tirelessly, and compromised at every turn. . . . The result: total defeat, no moral victories. So now we know what we didn't before: Making nice doesn't work. . . . We need to shame them, starting now."[7]

A similar anger was shared by Michael Brune, executive director of the Sierra Club, in a May 2010 commentary on CNN about the BP oil spill in the Gulf of Mexico: "Our country has huge solar power potential as well. We can also save more oil through simple efficiency measures than could be recovered by new drilling on our coastlines. This oil spill changes everything. We have hit rock-bottom in our fossil-fuel addiction. This tragedy should be a wake-up call."[8] At the same time, a rising number of informed yet over-committed college and university presidents are now clarifying their core messages and asking their leadership teams: What should we actually be *doing* when we are providing sustainability leadership on and beyond our campuses? What should our behaviors mean and our actions look like?

Glenn Cummings, deputy assistant secretary of the U.S. Department of Education, posed the same questions in his candid assessment of sustainability leadership among senior administrative officers: "As colleges and universities throughout the country accept their collective responsibility for educating the next generation in an idea loosely called 'sustainability,' the mission of becoming better stewards of the earth has expanded on American college campuses. . . . Nevertheless, the implementation methods used within institutions remain complicated, strewn with obstacles and weighted with risks. Leadership faces complex political, structural and personal barriers to significant change in the pursuit of sustainability."[9]

As many presidents and trustees have been discussing the need for immediate action plans, the Princeton Review has added a key question to its high-profile annual survey of students and parents. In 2008, the Review added its first "green" question, querying whether information about an institution's dedication to the environment would be useful in the data collection process. In the first year, 63 percent of respondents said yes, and in 2009, the total was 66 percent. The eventual publication, *The Princeton Review's Guide to 286 Green Colleges,* focused on success in three areas of the campus experience: providing students a healthy and sustainable quality of life, preparing students effectively for green jobs, and employing environmentally responsible institutional policies and practices.[10] With these three guidelines and the broader philosophical challenges posed by activists like McKibben and Brune as background, we offer the following group of summary best practices drawn from our own research and from chapters by various contributors to this book. These recommendations focus on institutionalizing sustainability thinking and decision making across the culture of a college or university while accounting for the realities of fluctuating budgets, administrative expertise, and student preferences.

Summary Best Practices for Presidents, Provosts, and Trustees

In developing any set of summary best practices related to sustainability action, a 2009 comment by John Tallmadge outlines the scope of the task: "Sustainability is not a problem, a condition, or a program; it's a way of life. . . . Think of sustainability as a type of infinite game, in which the goal is not to win (which would end the game), but to keep on playing forever. In practical terms, sustainability must always manifest itself in some place with some people; it always has a local, personal flavor. And because conditions and people change, sustainability always appears dynamic and evolving."[11] Or, as Bill McKibben adds, "The 'we' in this case [is] not rich white folks. If you look at the 25,000 pictures on our Flickr account, you'll see that most of them [are] poor, black, brown, Asian, and young—because that's what most of the world is. No need for vice presidents of big conservation groups to patronize them: shrimpers in Louisiana and women in burqas and priests in Orthodox churches and slumdwellers in Mombasa turned out to be completely capable of understanding the threat to the future."[12] The focus of this volume, thus, is to provide a comprehensive resource for presidents, trustees, provosts, and senior leadership that identifies specific challenges and provides action plans to address them. As the relationship between sustainability and higher education leadership becomes increasingly complex, these best practices address this dynamic.

1. *Hire a director of sustainability.* Perhaps surprisingly, there is still no consensus on whether the institutionalization of sustainability should be led by a single person with the title director or coordinator. AASHE does not yet maintain a count of how many of America's 4,100-plus higher education institutions employ a designated sustainability director, coordinator, or manager, but their firmest projection as of October 2009 was 300–800, with 500–600 individuals the most likely total.[13] These estimates were corroborated by the authors in numerous informal conversations with sustainability professionals in 2008 and 2009. Their perception was

Summary of Best Practices for Higher Education Leaders

1. Hire a director of sustainability.
2. Make public institutional commitments concerning sustainability.
3. Do not follow the money, lead it.
4. As president, advocate and promote.
5. Be pragmatic about LEED.
6. Challenge athletics to participate.
7. Find niches and market them.
8. Treat students as partners and stakeholders.

that whatever the specific number is—AASHE is considering a model to survey all directors and coordinators to determine their number and the common responsibilities within position descriptions—it is growing rapidly when judged by records of attendance at regional and national sustainability meetings and observation of national university job postings.

Laura Matson offers this assessment of the position and its evolution: "As institutions are increasingly realizing, having staff dedicated to the coordination of sustainability efforts is an institutional best practice. . . . Having an office or individual to coordinate sustainability initiatives and connect the relevant stakeholders helps to bridge the silos of higher education institutions. At the same time, it is important to realize that sustainability is not the responsibility solely of the dedicated staff—all campus community members can contribute."[14] This final point was affirmed by Howard Sacks, director of the Rural Life Center at Kenyon College, when he observed that it is a mistake to believe that "sustainability can be addressed by a single office or division of a college without investment by all constituencies."[15]

While a lack of consensus may linger, it is our belief that hiring a director of sustainability is the first best practice a university or college should implement in moving its agenda forward.

2. *Make public institutional commitments concerning sustainability.* This best practice generally accompanies the successful institutionalization of sustainability thinking. Whether or not a college or university has reached the point of hiring a director or coordinator of sustainability, there remains the need to engage the institution in the national conversation about sustainable goals and objectives and to challenge the school to make public its agenda continuously via web updates, presidential advocacies, and budget dedications. While joining AASHE, the

Presidents' Climate Commitment, and STARS have been among the highest-profile activities, they should not be perceived as the institution's total effort, especially for institutions seeking increased credibility for their sustainability values. Some colleges, such as those in the Eco League, confirm their commitments with specific language in their mission statements:

—"Green Mountain College prepares students for productive, caring, and fulfilling lives by taking the environment as the unifying theme underlying its academic and co-curricular programs."

—"College of the Atlantic enriches the liberal arts tradition through a distinctive educational philosophy—human ecology."

—"Northland College integrates liberal arts studies with an environmental emphasis, enabling those it serves to address the challenges of the future."

—"It is the mission of Prescott College to educate students of diverse ages and backgrounds to understand, thrive in, and enhance our world community and environment."[16]

Other institutions have joined groups like "Green Generation," a two-year initiative within the Earth Day Network culminating on the fortieth anniversary of Earth Day in 2010, or partnered with an organization such as the Sierra Student Coalition, a nationwide network of high school and college-aged students working to protect the environment.[17] Even these activities will not touch the lives of the full campus, however, so a growing number of higher education institutions are increasing employee accountability for sustainable objectives by integrating them into job descriptions and performance evaluations.[18] Julian Keniry, director of campus and community leadership for the National Wildlife Foundation's Campus Ecology program, agrees: "Putting performance goals and objectives into faculty and staff evaluations can help keep university employees cognizant of ongoing sustainability issues" as well as helping that institution look beyond small reforms in carbon usage to appreciating sustainability as an issue woven more broadly into the fabric of community life.[19]

At the same time, it is important to remember higher education sociologist Burton Clark's warning: "A large number of . . . universities, perhaps a majority, will not venture very far down the road of self-induced major change. All the more impressive are the feats of those universities that not only overcome their fear of failure before setting out on a journey of transformation, but also accomplish to a significant degree the second miracle of maintaining the will to change for a full decade and beyond."[20]

3. *Do not follow the money, lead it.* Structuring budgets to achieve sustainable solutions has proven difficult for several unrelated reasons. Beyond the "silo" impact of competing operational areas and the lack of incentives for some budget managers to try to save funds for sustainable goals, there are other hurdles. As Norb Dunkel noted in conversations with James Martin, there is still not a broadly accepted business plan understanding of the return on investments, thus causing some chief financial officers to turn toward quicker, simpler decisions such as funding the overdue deferred maintenance projects found on almost every campus. Additionally, some sustainability projects and personnel may be grant funded, causing not only their long-term stability to be uncertain but also, if the money has been derived from a partnership with a corporation or a government agency, possibly "foster[ing] social conservatism and weaken[ing] the legitimacy of social criticism and social engagement" on campus.[21]

Finally, there is the "hype factor." Higher education is not immune from pressures to fund the "politically sexy" programs described in an earlier chapter by Robert Weygand, the University of Rhode Island's current chief finance officer and a former U.S. congressman and lieutenant governor for the state. Familiar with much of the internal conflict that can shape a multi-hundred-million-dollar budget, Weygand cautions, "Budgeting during these highly contentious economic times has forced all higher education institutions to retool their financial analysis to determine the most worthy programs that warrant discretionary spending. We must not fall victim to pleas."[22]

One of the most successful approaches to efficient sustainability budgeting in recent years has been a simple one in concept. Advocated by Leith Sharp, founding director of the Harvard Green Campus Initiative, among others, the model stresses the importance of capturing cost savings from sustainability projects that have a shorter payback period and using those savings to fund projects without the same bottom-line advantages but significant sustainability benefits. This encourages colleges and universities to avoid tempting, low-hanging-fruit projects that would undercut the ability to finance later, larger, more ambitious ones.[23]

4. *As president, advocate and promote.* The benefits of signing the ACUPCC document are well recorded in terms of national press; however, observers are now noting whether a given institution moves forward from that point and distinguishes itself for continuing sustainability-related

programs and innovations. Whatever its decision, the institution's progress can be significantly shaped by the public and personal capital its chief executive officer is willing to spend in support of this agenda.

Stephen J. Nelson, the author of *Leaders in the Labyrinth: College Presidents and the Battleground of Creeds and Convictions,* contends that a "critical feature that determines success or failure for college presidents is their use, and sometimes abuse, of the bully pulpit. The bully pulpit is the weightiest possession of the college president. . . . Whenever [they] climb into the pulpit, they must do so not with the intent to please . . . but to edify, persuade, judge, and admonish."[24] Jo Ann Gora, president of Ball State University, was one of the twelve charter signatories of the ACUPCC in 2006. Continuing the momentum created early in her presidency, Gora and her leadership team broke ground in 2009 for the largest geothermal district energy system in the country, which when completed will heat and cool forty buildings on the Ball State campus, reducing the institution's current carbon footprint by almost 50 percent and saving $2 million annually.[25]

Serving as co-host, with AASHE, of the first Green Campus Exposition in 2009, Gora offered this view of the president's role as both advocate and promoter: "Most have read Covey's leadership treatises in which he makes the distinction between 'the important' and 'the urgent.' Sustainability is the perfect example of the 'important' issue that demands our attention in a quiet, persistent way. As presidents, if we do not attend to the 'important' issues, we never have the chance to make an impact and set the stage for transformational change. It is not enough for us simply to put out fires via 'urgent' issues; focusing on sustainability forces us all to think long-term and to effect real change on our campuses and in our students' minds."[26]

David Skorton, president of Cornell University, is another activist chief executive who understands the extra lengths that the president alone must travel to accomplish a sustainability agenda that is authentic on campus: "Because there are no real cost-effective solutions to achieve climate neutrality today, a strong emphasis on education and research, coupled with the willingness to make tough decisions now, will produce meaningful answers for tomorrow. . . . We will use our leadership opportunity to encourage more public and private investments in investigations that will yield . . . better approaches to the problems."[27] The depth to which presidents will personally engage in and promote their campus's sustainability vision carries a direct impact on institutionalizing sustainability thinking.

5. *Be pragmatic about LEED.* To this point, LEED certifications have been discussed in the contexts of both challenges and misconcep-

tions. In light of their increasing influence on sustainability resources and programming, a related best practice is also useful. To put it succinctly, institutional communities, before applying, should be candid with themselves about whether a certification will reflect mission, respect resources, and meet future needs.

The current environment for LEED decision making extends from single-family presidential dwellings to multi-campus construction plans easily achieving half a billion dollars in value. One architect specializing in sustainable design interviewed in the *Chronicle of Higher Education* admitted that the "vast majority" of her clients to date have decided not to pursue LEED certification,[28] yet the power and prestige of its brand continues to increase to the point that almost anyone on a campus with environmental awareness carries a formed opinion on one side or the other.

Whether it is a high-profile case, such as Santa Clara University's decision not to pursue LEED certification for its Commons on Kennedy Mall in 2006 because the process would have incurred $250,000 of registration, certification, and documentation paperwork as well as an additional $200,000 of LEED-mandated instrumentation to monitor the technologies being installed in the building,[29] or the decision by Bowdoin College to pursue, and earn, LEED certification for the first newly constructed ice arena in the nation in 2009,[30] LEED considerations and their growing impact must be addressed by campus sustainability managers in their budget, personnel, and program planning. Thus, these final decisions should incorporate rigorous assessments of mission, resources, and the college or university's strategic goals.

6. *Challenge athletics to participate.* Early in the development of this book, the director of sustainability at a university recognized nationally for its Division I sports teams and championships described university athletics programs and medical centers, collectively, as the "holy grail" for those attempting to achieve serious gains in sustainability institutionalization and innovation.[31] This best practice focuses on bringing senior athletics administrators more fully into the campus conversation with the intent to achieve higher profile and higher value sustainability commitments.

While one can point to the striking achievement in 2009 of LEED platinum certification for the University of Florida's Heavener Football Complex, the nation's first athletic facility to reach platinum status, described in detail in chapter 16 of this volume, and a few other LEED-certified, sports-related buildings (e.g., the Petersen Athletic Academic Center at the University of Nevada–Reno and the Recreation Center at the College of William and Mary), there is still not a comprehensive movement nationally to link sustainability with university athletic programs and infrastructure. Added to this, media attention remains more focused on individual press-worthy events such as the first carbon-neutral college football game between Florida and Florida State University in 2007.

In part, this is a reflection of the fact that, as former University of Michigan president James Duderstadt laments, "Big-time college athletics has little to do with the nature and objectives of the contemporary university. Instead, it is a commercial venture, aimed primarily at providing public entertainment for those beyond the campus."[32] Developing a set of sustainability best practices for athletic programs, while a daunting prospect, is still achievable, however. The CFO of one New England university recommends that the best way to get buy-in from this operational area is to "reassign energy expenses to the individual departments. This has proven to be an effective form of responsibility-centered management on our campus. Departments, including athletics, will be far more motivated to save energy and its related dollars if they see those costs as part of their budget. In our experience, athletics becomes more energy conscious as they see it is in their best financial interest."[33]

While an adjusted energy budget may not be a major coup, it is both a tangible step and a broader statement to one of the most important areas on campus necessary for sustainability buy-in. Ironically, one cohort often excluded from this conversation and its decision making are the athletes themselves. Authors of "The Green Athlete," who are competitors themselves, advise: "Forget trendy and forget the irony of being a 'green consumer.' We athletes make use of the earth in a way that no one else does, and so it's all the more important for us to take a realistic look around at how our . . . pursuits impact our environments."[34]

7. *Find niches and market them.* Since many sustainability initiatives are driven by a combination of student tuition and focused philanthropy, it has become increasingly important to identify niches that distinguish colleges and universities among a sea of competitors for both recruitment and fundraising purposes. *University Business* magazine reported in 2008 that incoming students were twice as likely to choose an undergraduate institution based on sustainability options than were students just three years previously. Almost 15 percent of incoming students are now taking environmental programs into consideration, and while some still view this as a small number, others are focusing instead on how the percentage of students has doubled in only thirty-six months.[35]

The University of Maine–Fort Kent, as one example, secured its sustainability niche in early 2008 when the trustees of the Maine University System approved the establishment of the Center for Rural Sustainable

Development on its campus. The center is a collaborative effort among UMFK and campuses in Machias and Presque Isle, and it will partner with local communities and economic development organizations to advance the concept that sustainable actions should be shaped by five integrated dimensions: human, economic, environmental, technological, and political.[36]

A different approach was taken by the University of Texas Health Science Center in its development of a series of green roofs that now serve as a chain of outdoor classrooms in downtown Houston. The roofs reduce storm water runoff and airborne pollutants while lowering university energy costs by insulating against cold and shielding against the summer sun.[37] Fifty miles outside of Las Vegas, another niche was developed through collaboration among several University of Nevada–Reno faculty members. Sensing the need for "boutique" herbs in the kitchens of casinos that attract more than 40 million visitors annually from almost every country in the world, the professors partnered with a group of Nevada farmers to launch "Niche Products for Direct Markets: A Sustainable Opportunity for Small Desert Farmers." As well as herbs familiar to most palates, the collaborative can deliver custom orders of such exotics as nepitella and lovage on twenty-four hours' notice.[38]

In sum, finding niches and marketing them are less critical in the scheme of institutional priorities than hiring a director of sustainability—but not that much less, considering the financial stability a new class of students in related majors can have on a director's budget. As the authors of "Sustainable Admissions" advise, "Environmentalism is very much on the minds of today's students. Those who are highly attuned to these issues will seek out the information available on colleges' sustainability efforts from the variety of sources mentioned here and many others, including student and faculty bloggers."[39] As many competitor colleges and universities push to identify and then publicize at least one sustainable program, service, or recognition to lure new students, some institutions are choosing to "direct market" their achievements through the production of their own campus green guides. The Planet Green website offers this advice: "Create a Campus Guide. . . . because with one click, you're halfway toward customizing a chlorine-free green guide (with recycled cardboard covers, of course) for your specific campus. Transportation, clothing, energy usage . . . it's all included."[40]

8. *Treat students as partners and stakeholders.* While superlative amenities, fresh technologies, and remarkable professors are the ingredients by which many assess their undergraduate years, colleges and universities of distinction continue to probe the experiences of their students beyond

residence comfort and online access and regularly act on what they discover. As the author of "Generation Green" acknowledges, "Once derided as tree-huggers, eco-friendly youth are now the nation's most powerful (and feared) voting bloc. . . . The math is not complicated. At 100 million strong, Millennials—those born roughly between 1980 and 2000—are the single largest generation of Americans, ever." Within twenty years, this group will constitute two of every five voters in the country, yet for these Millennials, the fight is not just about climate change; it is also about "creating green jobs and increasing national security by reducing dependence on foreign oil."[41]

Successful institutions interpret this final best practice as the need to provide and sustain an authentic learning experience amid a menu of synthetic life choices. As the authors of the book *Authenticity* warn, the steady virtualization of life has led to a deep yearning among Millennials for the authentic, as shallowly constructed experiences become more prevalent throughout social networking systems.[42] In response, an institutional mission and academic program focus centered on sustainable community and resource development presents a powerful objective for colleges and universities to pursue regarding future student needs and expectations.

Sustainability Leadership in 2020

This book closes with as many concerns about sustainability leadership as solutions, not because of a lack of thoughtful innovations by presidents, deans, and sustainability planning teams, but rather because of the enormity of the challenges facing their institutions and the communities that support and trust them. As the American public observed during the 2010 Gulf of Mexico oil spill, the 2008–2009 economic crisis, and even during Hurricane Katrina in 2005, it only takes "hours for things to go wrong."[43] Perhaps ironically, Bill McKibben, who also serves as a scholar-in-residence at Middlebury College, views various uncertainties about the future with guarded optimism because of the power of the same technologies mentioned above and undergraduates' "intrinsic understanding of the world's interconnectedness, fostered by Facebook and social networking."[44]

As sustainable lifestyle decisions from preschool to assisted living become more prevalent on and surrounding their campuses, presidents must face the question of how best to lead their communities going forward. In closing, we interviewed a final time some of the sustainability professionals with whom we had spoken early in this project to hear their summary thoughts on what the most important skill will be for higher education leaders to develop over the coming decade. Davis Bookhart, director of sustainability efforts at the Johns Hopkins

Closing Thoughts on Sustainability Leadership for Presidents

- Reframe expectations.
- Accept responsibility for the institution.
- Lead well.

University, believes that in ten years sustainability professionals will be coordinating "more mature programs characterized by a stronger emphasis on planning and structure, increasingly specialized staffing, and more concrete goals. Sustainability leaders will no longer be burdened with the task of continuously justifying a vision of sustainability; rather, they will be challenged even more intensely with the task of implementing it."[45] Aurora Lang Winslade, in a similar leadership position at the University of California–Santa Cruz, projects that by 2020 presidents and provosts will need to move beyond the easy, first-level goals in sustainability planning and budgeting and develop skills in "systems thinking and being able to integrate more complexity into strategic plans. They will also need more highly developed interpersonal and change management skills, including participatory decision-making and community-based social marketing."[46]

Mary Jo Maydew, chief financial officer at Mount Holyoke, contends that the most important skill for presidents will be a very basic one: to lead as if there is no tension between being environmentally responsible and being cost effective, because they are separate and achievable.[47] Paul Rowland, executive director of AASHE, who introduced the conversations in this volume in his foreword, provides a succinct, closing observation that the major challenge for sustainability leaders over the next decade will be to articulate how sustainability underpins effective education, and the most important characteristics defining successful presidents and administrative teams will be the ability and willingness to take responsibility for what their graduates do in the future.[48]

Our own view is captured most effectively in the comments of a recently retired university president in a new journal of "creative sustainability" published by Unity College, the small Eco League institution in Maine. In his article, "The Silver Lining in Forced Frugality," David Shi, formerly of Furman University, speaks to presidents and provosts of the future: "Living a simpler life does not mean . . . a rural homestead or a faddish preference for L. L. Bean boots. . . . Rather, it entails a daily ordering of priorities so as to distinguish between the necessary and superfluous, the useful and wasteful, the beautiful and the vulgar. . . . Perhaps the painful recession will provoke at least some of us to reassess our priorities." Shi concludes with the observation that "although often buffeted by forces beyond our control, most of us have choices: We can keep yearning for more, or we can resolve to be content with less. Choose well."[49] While Shi ends with "choose well," we would close, for those presidents and trustees still searching for big answers, with an even simpler ten-year plan: Reframe expectations, accept responsibility, and lead well.

Notes

Chapter 1. The Sustainable University

1. Anthony Cortese, telephone interview with James Martin, 11 February 2010.
2. Geoff Chase, email to James Martin, 12 January 2010.
3. Anthony Cortese, "The Core Mission of Higher Education: Creating a Thriving, Civil and Sustainable Society" (paper presented at the ACPA Sustainability Institute, Harvard University, Cambridge, MA, 13 June 2000).
4. Geoff Chase, email to James Martin, 12 August 2009.
5. Aurora Winslade, email to James Martin, 8 August 2009.
6. Elia Powers, "Climate Change and the University," *Inside Higher Ed,* 1 May 2007, www.insidehighered.com/news/2007/05/01/climate.
7. Harry Wideman, "Notice: Human Hands at Work," *Northland College Magazine* (Summer 2009): 15.
8. Peter Bardaglio, "STARS Is the New Game in Town," *Greentree Gazette* (July/Aug 2009): 10.
9. Ibid., 12.
10. Ibid., 10.
11. Elia Powers, "Comparing Environmental Data," *Inside Higher Ed,* 2 April 2008, www.insidehighered.com/news/2008/04/02/sustainable.
12. Bardaglio, "STARS," 12.
13. Norb Dunkel, email to James Martin, 7 August 2009.
14. Aurora Winslade, email to James Martin, 8 August 2009.
15. Mel Wilson, "Investing in the Future," *Communications Review* 12, no. 1: 17.
16. Robert Weygand, email to James Martin, 12 August 2009.
17. Stephen Pelletier, "Sustainability: What Is the Trustee's Stake?" *Trusteeship* 16, no. 5 (2008): 10.
18. Ibid., 10, 11.
19. Geoff Chase, email to James Martin, 11 August 2009.
20. Laura Matson, email to James Martin, 6 August 2009.
21. Cortese, "Core Mission of Higher Education," 4.
22. Mitchell Thomashow, email to James Martin, 4 August 2009.

23. Alex Steffen, "It's Not Just Carbon, Stupid," *Wired* 16, no. 6 (June 2008): 165.

24. Scott Carlson, "Cost and Red Tape Hamper Colleges' Efforts to Go Green," *Chronicle of Higher Education,* 11 April 2008, http://chronicle.com/article/CostRed-Tape-Hamper/14129.

25. Michael Baer, email to James Martin, 6 August 2009.

26. James Geary and Marco Visscher, "New Model Army," *Ode Magazine* 6, no. 4 (May 2008): 61.

27. Richard Holledge, "Staying the Distance," *Financial Times,* 16–17 February 2008, 1.

28. Laura Matson, email to James Martin, 9 August 2009.

29. Wayne Curtis, "A Cautionary Tale," *Preservation* 60, no. 1 (2008): 24.

30. Carlson, "Cost and Red Tape Hamper Colleges' Efforts."

31. Mitchell Thomashow, email to James Martin, 4 August 2009.

32. Carlson, "Cost and Red Tape Hamper Colleges' Efforts."

33. Ibid.

34. Dave Newport, email to James Martin, 5 August 2009.

35. Joel Garreau, *Radical Evolution: The Promise and Peril of Enhancing Our Minds, Our Bodies—and What It Means to Be Human* (New York: Doubleday, 2005), 10.

36. Hannah Fearn, "Architects Doubt Benefits of 'Greenwash Ecotecture,'" *Times Higher Education,* 17 December 2009, www.timeshighereducation.co.uk/story.asp?storyCode=409580§ioncode=26. See the Sheppard Robson website at www.sheppardrobson.com/practice.

37. Rebecca Smith, "A Consumer's Guide to Going Green," *Wall Street Journal,* 12 November 2007, http://online.wsj.com/public/article/SB119463269049588171.html.

38. Dave Newport, email to James Martin, 5 August 2009.

39. Tiffany Hsu, "Eco-Officers Are Moving into Executive Suites," *Los Angeles Times,* 30 December 2009, http://articles.latimes.com/2009/dec/30/business/la-fi-green-officers30-2009dec30.

Chapter 2. Promises Made and Promises Lost

1. Dianne Dumanoski, *The End of the Long Summer: Why We Must Remake Our Civilization to Survive on a Volatile Earth* (New York: Crown Publishers, 2009), 21.

2. Ibid., 22.

3. Ibid., 2.

4. David W. Orr, *Earth in Mind: On Education, Environment, and the Human Prospect* (Washington, DC: Island Press, 1994), 1, 17.

5. Gary Rivlin, "Age of Riches: In Silicon Valley, Millionaires Who Don't Feel Rich," *New York Times,* 5 August 2007.

6. Bret Schulte, "Saving Earth, Saving Money," *U.S. News and World Report,* 1 October 2006, www.usnews.com/usnews/news/articles/061001/9qa.htm.

7. Ray C. Anderson, "Editorial: Earth Day, Then and Now," *Sustainability: The Journal of Record* 3, no. 2 (April 2010): 73–74.

8. Frank H. T. Rhodes, "Sustainability: the Ultimate Liberal Art," *Chronicle of Higher Education,* 20 October 2006, http://chronicle.com/article/Sustainability-the-Ultimate/29514.

9. U.S. Green Building Council homepage, www.usgbc.org.

10. Louis Menand, *The Market Place of Ideas* (New York: W.W. Norton, 2010), 17.

11. Derek Curtis Bok, *Universities and the Future of America* (Durham, NC: Duke University Press, 1990), 104–5.

12. See the Presidents' Climate Commitment webpage at www.presidentscli matecommitment.org.

13. James Buizer, in a conversation with Anthony Cortese, July 2010.

14. Michael Crow, "Towards Institutional Innovations in America's Colleges and Universities," *Trusteeship* 3, no. 18 (May/June 2010): 4.

15. Joel Garreau, email to James Martin, 18 September 2010.

16. Glen Cummings, "The Leadership Factor: Implementing Sustainability in Higher Education," 6 July 2010, www.presidentsclimatecommitment.org/pcc/newslet ter/022.html.

17. Anthony Cortese, Georges Dyer, and Michelle Dyer, "Leading Profound Change: A Resource for Presidents and Chancellors of the ACUPCC" (Boston, MA: ACUPCC, July 2009). www.presidentsclimatecommitment.org/files/documents/Leading _Profound_Change.pdf.

18. Richard Cook, private communication with Anthony Cortese, April 2010.

Chapter 3. Trends, Skills, and Strategies to Catalyze Sustainability across Institutions

1. Hart Research Associates, "Raising the Bar: Employers' Views on College Learning in the Wake of the Economic Downturn: A Survey among Employers Conducted on Behalf of the Association of American Colleges and Universities," www.aacu.org/leap/documents/2009_EmployerSurvey.pdf.

2. Peter Blaze Corcoran and E. J. Wals Arjen, eds. *Higher Education and the Challenge of Sustainability: Problematics, Promise, and Practice* (Dordrecht: Kluwer Academic Publishers, 2004), 3–6.

3. For strategic plans, see Association for the Advancement of Sustainability in Higher Education, "Strategic Plans that Incorporate Sustainability," www.aashe.org /resources/strategic_plans.php. For master plans, see "Master Plans that Incorporate Sustainability" at the same web address.

4. Association for the Advancement of Sustainability in Higher Education, "Higher Education Sustainability Position and Salary Survey," www.aashe.org/publi cations/surveys.

5. National Environmental Education Foundation, "The Business Case for Environmental and Sustainability Employee Education," February 2010, www.neefusa .org/business/index.htm.

6. Magdalena Svanström, Francisco Lozano, and Debra Rowe, "Learning Outcomes for Sustainable Development in Higher Education," *International Journal of Sustainability in Higher Education* 9, no. 3 (2008): 339–51.

7. Evergreen State College, "Evergreen Earns Top Acclaim in the Princeton Review's Green Guide," 28 April 2010, www.evergreen.edu/news/archive/2010/04 /princetonreview.htm.

8. Ben Barlow, *Financing Sustainability on Campus* (Washington, DC: National Association of College and University Business Officers, 2009).

9. Association for the Advancement of Sustainability in Higher Education, "Higher Education Sustainability Officer Position and Salary Survey," January 2008, www.aashe.org/documents/resources/pdf/sustainability_officer_survey_2008.pdf.

10. Julie Garrett, "Sustainability Savings Pay for Future Efforts," *Community College Times,* 6 July 2008, www.communitycollegetimes.com/article.cfm?ArticleId=1059.

11. UBC Campus Sustainability Office, "UBC Case Study: Institutionalizing Sustainability" (University of British Columbia, 2008–2009), 27. www.sustain.ubc.ca/campus-sustainability/resource-exchange/case-studies.

12. The Princeton Review, "2010 College Hopes & Worries Survey Report" (2010): 4, www.princetonreview.com/uploadedFiles/Test_Preparation/Hopes_and_Worries/HopeAndWorries_Full%20Report.pdf.

13. Second Nature, "Advancing Green Building in Higher Education," http://secondnature.org/AGB.html.

14. Tony Cortese in an email message to authors, 6 August 2010.

15. Higher Education Associations Sustainability Consortium (HEASC) informational brochure, www2.aashe.org/heasc/index.php.

16. Higher Education Associations Sustainability Consortium, "Resource Center," www2.aashe.org/heasc/resources.php.

17. Rolf Jucker, "'Sustainability? Never Heard of It!': Some Basics We Shouldn't Ignore When Engaging in Education for Sustainability," *International Journal of Sustainability in Higher Education* 3, no. 1 (2002): 8–18.

18. Karla Hignite, "The Peaceable Workplace," *Business Officer Magazine* 22, no. 1 (July/August 2010), www.nacubo.org/Business_Officer_Magazine/Current_Issue/JulyAugust_2010/The_Peaceable_Workplace.html.

19. Debra Rowe, "Building Political Acceptance for Sustainability: Degree Requirements for All Graduates," in *Sustainability on Campus: Stories and Strategies for Change,* ed. Peggy Barlett and Geoffrey Chase (Cambridge, MA: MIT Press, 2004), 139–58.

20. Doug McKenzie-Mohr, *Fostering Sustainable Behavior: Community-Based Social Marketing* (Gabriola Island, BC: New Society Publishers, 1999). Available at: www.cbsm.com/pages/guide/preface.

21. Association for the Advancement of Sustainability in Higher Education, "Student Sustainability Educators Program," www.aashe.org/resources/peer2peer.php.

22. See the University of California–Santa Barbara LabRATS website at http://sustainability.ucsb.edu/LARS/purpose.php.

23. See the learning sessions for the Media Strategies for Sustainability conference at http://sites.google.com/site/campussustainmedia/learning-sessions.

24. American College & University Presidents' Climate Commitment, "Academic Guidance Document," www.presidentsclimatecommitment.org/resources/guidance-documents/academic.

25. Colby et al., *Educating Citizens: Preparing America's Undergraduates for Lives of Moral and Civic Responsibility* (San Francisco: Jossey-Bass, 2003).

26. Svanström, Lozano, and Rowe, "Learning Outcomes for Sustainable Development."

27. Washington Center, "Curricular Initiatives: Curriculum for the Bioregion," www.evergreen.edu/washcenter/project.asp?pid=62.

28. Debra Rowe, "Environmental Literacy and Sustainability as Core Requirements: Success Stories and Models," in *Teaching Sustainability at Universities: Towards Curriculum Greening,* ed. Walter Leal Filho (New York: Peter Lang, 2002). Available at: www.ncseonline.org/EFS/DebraRowe.pdf.

29. Association for the Advancement of Sustainability in Higher Education, "STARS: Sustainability Tracking Assessment & Rating System," http://stars.aashe.org.

30. Daniel Altman, "A Nobel That Bridges Economics and Psychology," *New York Times,* 10 October 2002.

31. Tim Jackson, *Prosperity without Growth: Economics for a Finite Planet* (London: Earthscan, 2009).

32. Joseph Stiglitz, *Freefall: America, Free Markets, and the Sinking of the World Economy* (New York: W.W. Norton, 2010).

33. "The U.S. Partnership for Education for Sustainable Development," http://usp.umfglobal.org/main/show_passage/51.

34. ACPA College Student Educators International, "ACPA's Presidential Taskforce on Sustainability: Change Agent Abilities Required to Help Create a Sustainable Future," www.myacpa.org/task-force/sustainability/docs/Change_Agent_Skills.pdf.

35. Center for Nonviolent Communication, "What Is NVC?" www.cnvc.org/en/about/what-is-nvc.html.

36. Debra Rowe, "Education and Action for a Sustainable Future" (keynote address, statewide sustainability conference, Kansas State University, Manhattan, KS, 29 January 2010).

37. Nicholas Wade, "We May Be Born with an Urge to Help," *New York Times,* 30 November 2009.

38. Frances Moore Lappé (lecture, Education for Sustainable Living Program, University of California–Santa Cruz, April 2007).

39. Penn State website, "Positive Psychology," www.ppc.sas.upenn.edu.

Chapter 4. Measuring Campus Sustainability Performance

1. Higher Education Associations Sustainability Consortium (HEASC), "Call for a System for Assessing & Comparing Progress in Campus Sustainability," www.aashe.org/files/documents/STARS/HEASCcall.pdf.

2. STARS 1.0 Technical Manual (Association for the Advancement of Sustainability in Higher Education, 2010), 2. Available at: www.aashe.org/files/documents/STARS/STARS_1.0.1_Technical_Manual.pdf.

3. Gro Harlem Brundtland, ed., *Report of the World Commission on Environment and Development: Our Common Future* (New York: United Nations, 1987). Available at: www.un-documents.net/wced-ocf.htm.

4. SAM Indexes, "Dow Jones Sustainability Indexes," www.sustainability-index.com.

5. Princeton Review 2009, "College Hopes & Worries Survey" Findings, www.princetonreview.com/uploadedFiles/Test_Preparation/Hopes_and_Worries/college_hopes_worries_details.pdf.

6. John H. Pryor, et al., *The American Freshman: National Norms Fall 2008* (Los Angeles: Higher Education Research Institute, Graduate School of Education and Information Studies, University of California, Los Angeles, 2009).

7. Kyle C. Murphy, "Evaluating the Sustainability Tracking, Assessment, and Rating System (STARS) at the Evergreen State College" (master's thesis, Evergreen State College, 2009).

8. STARS 1.0 Technical Manual, 1.

Chapter 5. Institutionalizing Sustainability

1. David Chrislip, *The Collaborative Leadership Fieldbook: A Guide for Citizens and Civic Leaders* (San Francisco: Jossey-Bass, 2002).

2. Quoted in Anne Underwood, "Green, Greener, Greenest: Many Universities Are Finding New Ways to Live and Learn in an Effort to Be Environmentally Friendly," *Newsweek,* 18 August 2008.

3. Stuart Hart, *Capitalism at the Crossroads: Aligning Business, Earth, and Humanity,* 2nd ed. (Upper Saddle River, NJ: Wharton School Publishing, 2007), 113.

4. University of North Carolina Tomorrow Commission, Final Report, December 2007, www.northcarolina.edu/nctomorrow/reports/commission/Final_Report.pdf.

Chapter 6. Sustainability

Epigraphs. The first is from a list developed by emrgnc, a "research methodology used to develop the theory and practice of societal emergence," available at www.emrgnc.com.au. The second, a quote from Justice Potter Stewart, is from *Jacobellis v. Ohio* 378 U.S. 184 (1964).

1. This model is known as an Environmental Kuznets Curve (EKC), where the x-axis shows wealth and the y-axis shows pollution. In many developed nations, a curve develops demonstrating that wealth and pollution rise at a similar rate until a level of wealth is achieved where pollution levels peak and then begin to decline.

2. *Process of preparation of the Environmental Perspective to the Year 2000 and Beyond,* UN General Assembly Resolution 38/161, 19 December 1983, www.un.org/documents/ga/res/38/a38r161.htm.

3. *Report of the World Commission on Environment and Development,* UN General Assembly Resolution 42/187, 11 December 1987, www.un.org/documents/ga/res/42/ares42-187.htm.

4. Andrew W. Savitz and Karl Weber, *The Triple Bottom Line: How Today's Best-Run Companies Are Achieving Economic, Social, and Environmental Success—and How You Can Too* (San Francisco: Jossey-Bass, 2006), 22.

5. From the introduction to the European Commission's vision of sustainable development, http://ec.europa.eu/environment/eussd/.

6. From the Emory University Office of Sustainability Initiatives, http://sustainability.emory.edu.

7. "The Baltimore Sustainability Plan," April 2009, 27, www.baltimorecity.gov/LinkClick.aspx?fileticket=DtRcjL%2fIBcE%3d&tabid=128.

8. Ibid., 8.

9. Ken Wilber, founder of the Integral Institute, as quoted on the emrgnc website at www.emrgnc.com.au/definitions.htm.

10. Paul Hawken, *Blessed Unrest: How the Largest Social Movement in History Is Restoring Grace, Justice, and Beauty to the World* (New York: Penguin 2008).

11. Australian Government's *National Strategy for Ecological Sustainable Development,* www.environment.gov.au/esd/.

12. Peter Woodward, Shell Workshop, as quoted on the emrgnc website at www.emrgnc.com.au/definitions.htm.

Chapter 7. Sustainable Citizenship

Epigraph. Earth Charter Initiative, Preamble to *The Earth Charter*, www.earthcharter inaction.org.

1. Tom Atlee with Rosa Zubizarreta, *The Tao of Democracy: Using Co-Intelligence to Create a World That Works for All* (North Charleston, SC: Imprint Books, 2003).

2. Wuppertal Institute, "Prism of Sustainability," www.foeeurope.org/sustainability /sustain/t-content-prism.htm.

Chapter 8. Sustainability and the Presidency

1. Mission statement of the Disciplinary Associations Networks for Sustainability (DANS), www2.aashe.org/dans/about.php.

2. From the Higher Education Associations Sustainability Consortium (HEASC) home page, www2.aashe.org/heasc.

3. UNESCO 2003, United Nations Decade of Education for Sustainable Development (January 2005–December 2014), *Framework for a Draft International Implementation Scheme,* 4.

Chapter 9. Not So Fast

1. Conscious Wave, Inc., "LOHAS Background," *Lifestyles of Health and Sustainability (LOHAS) Online,* www.lohas.com/about.html.

2. GreenBiz Staff, "Wal-Mart Pledges to Sell 100 Million Compact Fluorescents in '07," *GreenBiz.com,* 29 November 2006, www.greenbiz.com/news/2006/11 /30/wal-mart-pledges-sell-100-million-compact-fluorescents-07.

3. American College & University Presidents' Climate Commitment, www .presidentslimatecommitment.org/html/commitment.php.

4. University of Wyoming Campus Sustainability Committee, "Greenhouse Gas Emissions Inventory for the University of Wyoming, Update Fiscal Year 2008," University of Wyoming, http://uwyo.edu/sustainability/pcc.asp.

5. Dustin Bleizeffer, "Wyo Coal Miners Dig Record Volume in 2008," *Wyoming Energy Journal: Coal* (2009): 6–9.

6. The National Coal Council, *The Urgency of Sustainable Coal: Executive Summary,* report prepared for the U.S. Secretary of Energy, May 2008, Library of Congress Catalog # 2008928663.

7. William L. Sigmon, "The Lure of Ultra-Supercritical: Exploring the Future of Coal-Burning." *Energybiz* 5, no. 5 (September–October 2008): 90–91.

8. Martin Rosenberg, "Dominion's Multifaceted Empire." *Energybiz* 5, no. 5 (September-October 2008): 104.

9. Michael Shellenberger and Ted Nordhaus, "The Death of Environmentalism: Global Warming Politics in a Post-Environmental World" (essay, Environmental Grantmakers Association, 29 April 2004).

10. Carol Frost, "UW's Role in Carbon Sequestration in Wyoming" (presentation, University of Wyoming Board of Trustees meeting, Laramie, Wyoming, 6 March 2009).

11. Delissa Hayano, "Guarding the Viability of Coal and Coal-Fired Power Plants: A Road Map for Wyoming's Cradle to Grave Regulation of Geologic CO_2 Sequestration," *Wyoming Law Review* 139 (2009): 139–41.

12. Mark Northam, email to authors, 4 August 2009.

13. Michael Pollen, interview by *Yale Environment 360,* 26 June 2008, http://e360.yale.edu/content/feature.msp?id=2031.

14. "Domestic Energy Policy: Our Economic Future" (panel, Jackson Hole Policy Institute 08: Senior Executive Energy Summit, Jackson Hole, Wyoming, 12 November 2008).

15. Shellenberger and Nordhaus, "Death of Environmentalism," 6.

16. Daniel M. Kammen, "The Rise of Renewable Energy." *Scientific American* 295, no. 3 (September 2006): 63.

17. Ibid., 62, 65–66.

18. University of Wyoming School of Energy Resources, *First Annual Report,* prepared for the Joint Minerals, Business and Economic Development Interim Committee, Joint Appropriations Interim Committee, and the Joint Education Interim Committee, October 2006, www.uwyo.edu/sersupport/docs/2006report.pdf.

19. University of Wyoming School of Energy Resources, *2008 Annual Report,* prepared for the Joint Minerals, Business and Economic Development Interim Committee, Joint Appropriations Interim Committee, and the Joint Education Interim Committee, October 2008, www.uwyo.edu/sersupport/docs/SER%20Annual%20Rpt%202008%20Final.pdf.

20. Mike Chesser, "Leadership and Learning: Driving toward Unprecedented Excellence," *Energybiz* 5, no. 5 (September–October 2008): 24.

21. Michael F. Maniates, "Individualization: Plant a Tree, Buy a Bike, Save the World?" *Global Environmental Politics* 1, no. 3 (August 2001): 33.

22. Ibid.

23. Shellenberger and Nordhaus, "Death of Environmentalism," 6.

24. Maniates, "Individualization," 33.

25. Greg Havens, Perry Chapman, and Bryan Irwin, "The Role of Sustainability in Campus Planning: A New England University Builds on the Land Grant Tradition," *New England Journal of Higher Education* 23, no. 2 (Fall 2008): 28.

26. Ibid., 29.

27. Donella H. Meadows et al., *Limits to Growth: The 30-Year Update* (White River Junction, VT: Chelsea Green Publishing Company, 2004), xvi.

28. Associated Press, "Obama Official: Ocean Wind Could Replace Coal," *msn.com,* 6 April 2009, www.msnbc.com/id/30072485.

29. Dave Freudenthal, telephone conversation with authors, 2 September 2009.

Chapter 10. The Importance of Sustainability in the Community College Setting

1. David Orr, *Ecological Literacy: Education and the Transition to a Postmodern World* (Albany: State University of New York Press, 1992), 101.

2. Margaret Robinson, quoting David Orr, in an email to Mary Spilde, 16 March 2009.

3. Bill McDonough and Michael Braungart, *Cradle to Cradle: Remaking the Way We Make Things* (New York: North Point Press, 2002), 123.

4. Mindy Feldbaum with Hollyce States, *Going Green: The Vital Role of Community Colleges in Building a Sustainable Future and Green Workforce,* 2008. Available at: www.greenforall.org/resources/going-green-the-vital-role-of-community-colleges-in-building-a-sustainable-future-and-green-workforce.htm.

5. James L. Elder, "Higher Education and the Clean Energy, Green Economy," *EDUCAUSE Review* 44, no. 6 (November–December 2009): 108–9.

6. Margaret J. Wheatley, *Leadership and the New Science: Discovering Order in a Chaotic World* (San Francisco: Berrett-Koehler Publishers, 1999).

Chapter 11. Sustainability, Leadership, and the Role of the Chief Academic Officer

1. The United Nations World Commission on Environment and Development, *Our Common Future* (New York: Oxford University Press, 1987).

2. David Orr, *Earth in Mind: On Education, Environment, and the Human Prospect* (Washington, DC: Island Press, 1993).

3. U.S. Department of Education, National Center for Education Statistics. *Digest of Education Statistics, 2008* (NCES 2009-020), http://nces.ed.gov/fastfacts/display.asp?id=98.

4. Karl E. Weick, "Educational Organizations as Loosely Coupled Systems," *Administrative Science Quarterly* 21, no. 1 (March 1976): 1–19.

5. Telephone conversation with Nancy Marlin, 4 August 2009.

6. John Tallmadge, *The Cincinnati Arch: Learning from Nature in the City* (Athens: University of Georgia Press, 2004).

7. "Curriculum for the Bioregion" Initiative, Sustainability Learning Outcomes. The Washington Center for Improving the Quality of Undergraduate Education, 2008, www.evergreen.edu/washcenter/project.asp?pid=62.

8. See AASHE Vision, Mission, and Goals for 2011, www.aashe.org/about/aashe-mission-vision-goals.

9. See, e.g., Arizona State University at http://schoolofsustainability.asu.edu. See Anthony D. Cortese, "Education for Sustainability: The University as a Model of Sustainability" (1999), http://secondnature.org/pdf/snwritings/articles/univmodel.pdf.

10. Information for this section was obtained from the four colleges' websites: Green Mountain, www.greenmtn.edu/academics/ela.aspx; Unity College, www.unity.edu/Academic/DistinctivePrograms/TheCoreCirriculum/TheCoreCirriculum.aspx; College of the Atlantic, www.coa.edu; Northland College, www.northland.edu.

11. George D. Kuh, *High-Impact Educational Practices: What They Are, Who Has Access to Them, and Why They Matter* (Washington, DC: Association of American Colleges and Universities, 2008), 17.

12. John Tallmadge, as quoted in: "What Is Creative Sustainability?" *Hawk and Handsaw: The Journal of Creative Sustainability* 2 (2009): 4.

13. Laura B. DeLind and Terry Link, "Place as the Nexus of a Sustainable Future: A Course for All of Us," in *Sustainability on Campus: Stories and Strategies for Change,* ed. Peggy Barlett and Geoff Chase (Cambridge, MA: MIT Press, 2004), 121–37, 129.

14. Larry A. Braskamp and Jon F. Wergin, "Forming New Social Partnerships," in *The Responsive University: Restructuring for High Performance,* ed. William G. Tierney (Baltimore: Johns Hopkins University Press, 1998), 62–71.

15. L. D. Camblin Jr. and J. A. Steger, "Rethinking Faculty Development," *Journal of Higher Education* 39, no. 1 (2000): 1–18.

16. Parker Palmer, *The Courage to Teach: Exploring the Inner Landscape of a Teacher's Life* (San Francisco: Jossey-Bass, 1998). See also J. F. Wergin, E. J. Mason,

and P. J. Munson, "The Practice of Faculty Development: An Experience-Derived Model," *Journal of Higher Education* 47, no. 3 (1976): 289–308.

17. The Tufts/Emory study in 2006–2007 explored the long-term impacts of the Tufts Environmental Literacy program, which had existed for five years, 12–16 years earlier. Though some faculty had retired or moved elsewhere, 57 percent of past participants responded to requests for information. The Piedmont Project at that time was five years old, and 83 percent of past participants took part in the study. See Peggy Barlett and Anne Rappaport, "Long-Term Impacts of Faculty Development Programs: The Experience of TELI and Piedmont," *College Teaching* 27, no 2 (2009): 73–82.

18. Barlett and Chase, *Sustainability on Campus*. See also Geoff Chase and Paul Rowland, "The Ponderosa Project: Infusing Sustainability in the Curriculum," in *Sustainability on Campus,* 91–106.

19. Mitchell Thomashow, *Ecological Identity: Becoming a Reflective Environmentalist* (Cambridge, MA: MIT Press, 1995). See also Christopher Uhl, *Developing Ecological Consciousness: Path to a Sustainable World* (New York: Rowman and Littlefield, 2004).

20. Arri Eisen, Anne Hall, Tong Soon Lee, and Jack Zupko, "Teaching Water: Connecting across Disciplines and into Daily Life to Address Complex Societal Issues," *College Teaching* 57, no. 2 (2009): 99–104.

21. Alan Naditz, "The Green MBA: From Campus to Corporate," *Sustainability* 1, no. 3 (2008): 178–82.

22. Chase and Rowland, "Ponderosa Project."

Chapter 12. Greening the Endowment

1. For more information on Harvard's Advisory Committee on Shareholders Responsibility, see: www.hcs.harvard.edu/~gsc/guide/resources/gsc_structure.html#acsr.

2. For more information on Carleton College's committee, see http://apps.carleton.edu/governance/cric.

3. For more information on the University of Vermont's Socially Responsible Investment Working Group, see www.uvm.edu/~stffcncl/pdf/sriwg012110timeline.pdf.

4. For more information on these neighborhood development efforts, see www.nytimes.com/2009/03/25/realestate/commercial/25haven.html, and www.upenn.edu/pennnews/news/university-pennsylvania-announces-75-million-mixed-use-development-university-city.

5. Survey respondents included Amherst, Barnard, Bowdoin, Bucknell, Colby, Colgate, Davidson, Hamilton, Haverford, Macalester, Mount Holyoke, Swarthmore, Vassar, Wellesley, and Wesleyan.

6. See Ball State University's College Sustainability Report Card 2010 at www.greenreportcard.org/report-card-2010/schools/ball-state-university/surveys/endowment-survey.

7. See the University of Minnesota's Sustainability Report Card 2010 at www.greenreportcard.org/report-card-2010/schools/university-of-minnesota.

8. *The Yale Endowment 2009,* 16–18, www.yale.edu/investments/Yale_Endowment_09.pdf.

9. Social Investment Forum Foundation, *Investment Consultants and Responsible Investing: Current Practice and Outlook in the United States* (December 2009), 12. www.socialinvest.org/resources/pubs/documents/Investment_consultant.pdf.

10. Kyle Johnson, *Social Investing* (Boston: Cambridge Associates, 2007).

11. Social Investment Forum, *2007 Report on Socially Responsible Investing Trends in the United States*, www.socialinvest.org/resources/pubs/documents/FINALExecSummary_2007_SIF_Trends_wlinks.pdf.

12. Social Investment Forum Foundation, *Investment Consultants and Responsible Investing: Current Practice and Outlook in the United States* (December 2009), 1–2. www.socialinvest.org/resources/pubs/documents/Investment_consultant.pdf.

13. United Nations Principles for Responsible Investment, *Report on Progress 2009: A Review of Signatories Progress and Guidance on Implementation*, www.unpri.org/files/PRI%20Report%20On%20Progress%2009.pdf.

14. The 109 figure is as of 5 June 2010. See the list of signatories at www.unpri.org/signatories.

15. See the College Sustainability Report Card website at www.greenreportcard.org.

Chapter 13. Sustainability and Higher Education Architecture

1. Dorothy D. Park, quoted by Peter Bardaglio, in an interview with Scott Carlson, 12 August 2009.

2. Rick Couture, in an interview with Scott Carlson, 13 August 2009.

3. A number of organizations cite these figures for building consumption and emissions. See www.usgbc.org/DisplayPage.aspx?CMSPageID=1729 and www.wbdg.org/design/minimize_consumption.php.

4. Melissa Gallagher-Rogers, in an interview with Scott Carlson, 24 March 2008.

5. Jaimie Van Mourik, "Growing from Building-Centric to Campus-Wide: USGBC's New Portfolio Program" (presentation at the Society for College and University Planning annual conference, 21 July 2009).

6. Mireya Navarro, "Some Buildings Not Living Up to Green Label," *New York Times*, 30 August 2009, www.nytimes.com/2009/08/31/science/earth/31leed.html.

7. Scott Carlson, "Cost and Red Tape Hamper Colleges' Efforts to Go Green," *Chronicle of Higher Education*, 11 April 2008, http://chronicle.com/article/CostRed-Tape-Hamper-Co/14129/. See also Daniel Brook, "It's Way Too Easy Being Green: The Decidedly Dupable System for Rating a Building's Greenness," *Slate*, 26 December 2007, www.slate.com/id/2180862.

8. Carlson, "Cost and Red Tape."

9. Daniel Brook, "LEED Compliance Not Required for Designing Green Buildings: Constructing Buildings to the LEED Standard Can Conserve Energy and Materials—or Be Exploited for Promotional Gain," *Scientific American*, October 2008, www.scientificamerican.com/article.cfm?id=leed-compliance-not-required.

10. Rachel Gutter, in an interview with Scott Carlson, 9 March 2009.

11. Robert Koester, in an interview with Scott Carlson, 5 August 2009.

12. Derek Cunz, "Integrated Delivery: Forget about the Contract, Focus on the Process." *Design Intelligence*, 7 July 2009, www.di.net/articles/archive/3100.

13. Koester interview, 2009.

14. See also Rex Miller, "A Mindshift for Sustainability," *Design Intelligence*, 7 July 2009, www.di.net/articles/archive/a_mindshift_sustainability. Miller further states, "One key to sustainable design is the use of multi-disciplinary teams. However, the common approach . . . leaves most of the knowledge present on such teams

out of the strategic decision phase. Clive Thomas Cain, in his book *Profitable Partnering for Lean Construction,* reports that 80 percent of the cost and knowledge of a project resides in the specialty trades and vendors. However, the current system does not bring them in until all of the key design decisions have been made."

15. Davis Langdon, "The Cost of Green Revisited: Reexamining the Feasibility and Cost Impact of Sustainable Design in the Light of Increased Market Adoption," July 2007, www.davislangdon.com/USA/Research/ResearchFinder/2007-The-Cost-of-Green-Revisited.

16. Koester interview, 2009.

17. Richard Bowen, in an interview with Scott Carlson, 9 June 2009.

18. Scott Carlson, "A Reputation Cemented: Higher Education's Greenest Building," *Chronicle of Higher Education,* 8 April 2008, http://chronicle.com/blogPost/A-Reputation-Cemented-Higher/5100.

19. Bowen interview, 2009.

20. Paybacks and achievements listed by Richard Bowen in an email to Scott Carlson, 14 August 2009.

21. Jay Pearlman, in an interview with Scott Carlson, 24 February 2009.

22. Philip Parsons, in an interview with Scott Carlson, 20 March 2009.

23. Rodney Rose, *Buildings . . . The Gifts That Keep on Taking* (Alexandria, VA: APPA Center for Facilities Research, 2007), 62.

24. Scott Carlson, "As Campuses Crumble, Budgets Are Crunched," *Chronicle of Higher Education,* 23 May 2008, http://chronicle.com/article/As-Campuses-Crumble-Budget/6522.

25. Scott Carlson, "The $7-Billion Patch for Campus Maintenance," *Chronicle of Higher Education,* 30 January 2009, http://chronicle.com/article/The-7-Billion-Patch-for-Ca/1490.

26. James A. Willis, executive vice president for university support services, in an interview with Scott Carlson, 17 June 2009.

27. Scott Carlson, "Regular Checkups of Green Buildings Can Yield Millions in Savings," *Chronicle of Higher Education,* 12 October 2009, http://chronicle.com/article/Regular-Checkups-of-Green/48756.

28. Scott Carlson, "Ohio State U. May Put the Brakes on New Academic Space," *Chronicle of Higher Education,* 17 June 2010, http://chronicle.com/article/Ohio-State-U-May-Put-the/65984.

29. Parsons interview, 2009.

30. Scott Carlson, "Campus Officials Seek Building Efficiencies, One Square Foot at a Time," *Chronicle of Higher Education,* 17 April 2009, http://chronicle.com/article/Campus-Officials-Seek-Build/3292.

31. Scott Carlson, "After $74-Million and Counting, Frank Gehry's Library Opens at Princeton," *Chronicle of Higher Education,* 10 September 2008, http://chronicle.com/article/After-74-MillionCount/32435.

32. Scott Carlson, "Dazzling Designs, at a Price," *Chronicle of Higher Education,* 26 January 2001, http://chronicle.com/article/Dazzling-Designs-at-a-Price/7969.

33. Scott Carlson, "Push for Greener Buildings Gains Advocates, but When Is Green Really Green?" *Chronicle of Higher Education,* 29 June 2009, http://chronicle.com/article/When-Is-Green-Really-Green-/46955.

34. Koester interview, 2009.

35. Scott Carlson, "Oregon Universities Plan an Ambitiously Green Building in the Heart of Portland," *Chronicle of Higher Education,* 17 August 2009, http://chronicle.com/article/Oregon-Universities-Plan-an/48008. This article lists the cost and size of the building at 220,000 square feet for $90 million. Note that after the article was published, the building's planners scaled down the size and cost of the building amid the economic recession.

36. Living Building standards are available on the website for the International Living Building Institute: http://ilbi.org/the-standard.

37. The Living Building executive summary and matrix are available on the website for the International Living Building Institute: http://ilbi.org/resources/research/financial-study.

38. Jay Kenton, in an interview with Scott Carlson, 8 August 2009.

39. Wim Wiewel, email to James Martin, 9 September 2010.

40. As of this writing, the building has been redesigned once again to reduce interior space and allow for more roof space.

41. Clark Brockman, in an interview with Scott Carlson, 9 August 2009.

42. Brockman interview, 2009.

43. Kenton interview, 2009.

Chapter 14. Sustainable Campus Housing

1. Maruja Torres-Antonini and Norbert Dunkel, "Green Residence Halls Are Here: Current Trends in Sustainable Campus Housing," *Journal of College and University Student Housing* 36, no. 1 (2009): 14.

2. Ibid.

3. David Orr, "Green Residence Halls Are Here: Current Trends in Sustainable Campus Housing," *Journal of College and University Student Housing* 36, no. 1 (2009): 21.

4. Joe L. Kincheloe, *Critical Pedagogy* (New York: Peter Lang, 2008), 1–2.

5. University of Plymouth, "Centre for Sustainable Futures," http://csf.plymouth.ac.uk/?q=book/export/html1/601.

6. Emory University, "Make Your Personal Sustainability Pledge," http://sustainability.emory.edu.

7. Emory University, "Piedmont Project," http://sustainability.emory.edu/page/1021/Piedmont-Project.

8. Emory University, "Emory as Place," http://sustainability.emory.edu/page/1032/Service-Learning-Opportunities.

9. John McLain, "Prisons Turning Green: Evergreen Graduates, Students, and Faculty Put Their Stamp on Corrections Work," www.evergreen.edu/magazine/2009spring/greenprisons.htm.

10. Ibid.

11. Ibid.

12. Cornell University Campus Life, "Campus Life: Contributing to a Sustainable Environment," www.campuslife.cornell.edu/campuslife/housing/upload/sustainability-brochure-08-09-web-version.pdf.

13. Debra Hill, "Smart Home Dorm Opens Its Doors: Ten-Person Residence Is a Living Laboratory," www.dukenews.duke.edu/2007/11/smarthouse.html.

14. Laurence O'Sullivan, "Carbon Dioxide and Global Warming," http://pollution-control.suite101.com/article.cfm/carbon_dioxide_and_global_warming.

15. Energy Saves, "Passive Solar Home Design," www.energysavers.gov/your_home/designing_remodeling/index.cfm/mytopic=10250.

16. Lynn Deninger and John Swift, "Integrated Design: A Sustainable Mindset for Residence Halls," *Journal of College and University Student Housing* 36, no. 1 (2009): 53.

17. 2007 Buildings Energy Data Book, "How We Use Energy in Our Home," www1.eere.energy.you/consumer/tips/home_energy.html.

18. See Environmentalism, http://environmentalism.suite101.com/article.cfm/cutting_carbon_footprints.

19. University of Tennessee, "Turn Down the Heat, and Still Stay Warm," www.utk.edu/features/switch/staywarm.shtml.

20. American Council for an Energy Efficient Economy, "Consumer Guide to Home Energy Savings: Online Guide," www.aceee.org/consumerguide/index.htm.

21. Plymouth State University, "PSU Students Do It in the Dark, Save Big On Energy," http://greenup.blogs.plymouth.edu/2007/12/07/psu-students-do-it-in-the-dark-save-big-on-energy.

22. H. A. Belch, M. E. Wilson, and N. W. Dunkel, "Cultures of Success: Recruiting and Retaining New Live-In Residence Life Professionals," *Journal of College Student Development* 27, no. 2 (2009): 176–93.

Chapter 15. Food for Thought

1. See the cover for *Time* magazine, 2 March 2007.

2. For more information on Kenyon College's local food initiative, Food for Thought, visit http://rurallife.kenyon.edu.

Chapter 16. University Athletics and Sustainability

1. See Kate Galbraith, "Ranking Universities by 'Greenness,'" *New York Times*, Green blog, 20 August 2009, http://greeninc.blogs.nytimes.com/2009/08/20/ranking-universities-by-greenness. See also Francesca Di Meglio, "Building Green Classrooms," *Business Week*, 9 September 2008, www.businessweek.com/bschools/content/sep2008/bs2008099_035677.htm.

2. See the AASHE website at www.aashe.org/resources/resource_center.php.

3. Mark McSherry, 2009 Collegiate Athletic Department Sustainability Survey Report, June 2009, www.aashe.org/files/documents/resources/2009-Collegiate-Athletic-Department-Sustainability-Survey.pdf.

4. For NCAA 2008 Attendance Records, see http://web1.ncaa.org/web_files/stats/football_records/DI/2009/2008Attendance.pdf.

5. Julian Dautremont Smith, "How Do Campus Sustainability Initiatives Affect College Submissions?" AASHE Campus Sustainability Perspectives Blog, 2 March 2009, www.aashe.org/blog/how-do-campus-sustainability-initiatives-affect-college-admissions.

6. For information on green colleges from the Princeton Review, see www.princetonreview.com/green.aspx. See also "*Sierra Magazine*'s College Rankings: A Comprehensive Guide to the Most Eco-Enlightened U.S. Universities," www.sierraclub.org/sierra/200909/coolschools/default.aspx.

7. For the College Sustainability Report Card, see www.greenreportcard.org/report-card-2010/methodology.

8. Kate Zernike, "Green, Greener, Greenest," *New York Times,* 27 July 2008, www.nytimes.com/2008/07/27/education/edlife/27green.html.

9. Niles Barnes, "AASHE Passes New Milestones in Time for Earth Day," AASHE Campus Sustainability Perspectives Blog, 21 April 2010, www.aashe.org/blog/aashe-passes-new-milestones-time-earth-day.

10. See McSherry, Athletic Department Sustainability Survey Report.

11. "Sports and Sustainability," *Inside Higher Ed,* 30 July 2009, www.insidehighered.com/news/2009/07/30/sustainable; "Behind the News," *University Business,* September 2009, www.universitybusiness.com/viewarticle.aspx?articleid=1368&p=4.

12. "'Swamp' Goes Green with Help of Carbon Neutral Plan," University of Florida News, 20 November 2007, http://news.ufl.edu/2007/11/20/green-swamp.

13. Anne Prizzia, telephone interview with DeLongpré Johnston, 29 September 2009.

14. Nathan Crabbe, "Orange, Blue, and Green," *Gainesville Sun,* Gatorsports.com, 22 May 2009, www.gatorsports.com/article/20090522/ARTICLES/905229972?Title=Orange-Blue-and-green.

15. Press release, "RDG Project Achieves LEED Platinum!" June 2009, www.rdgusa.com/news/archives/2009/06/rdg_project_to_achieve_leed_pl_1.php.

16. Press release, "Zero-Waste Program Set for Football Games," 5 August 2008, www.cubuffs.com/ViewArticle.dbml?DB_LANG=C&DB_OEM_ID=600&ATCLID=1549817&SPID=255&SPSID=3843.

17. For information on UC Davis's zero-waste Aggie Stadium, see http://r4.ucdavis.edu/programs/zeroWaste/aggieStadium.php.

18. For information on Seal of Testing Assurance composting, visit the Composting Council's website at www.compostingcouncil.org/programs/sta.

19. "Ralphie's Green Stampede to Cover All CU Competitive Sports This Year," www.cubuffs.com/ViewArticle.dbml?DB_LANG=C&DB_OEM_ID=600&ATCLID=204782686&SPID=274&SPSID=4457.

20. Andrew Garnder, "Middlebury Athletics Adopts Green Mission," 8 June 2009, www.middlebury.edu/athletics/about/generalnews/node/112741.

21. Michelle Brutlag Hosick, "Study Says College Athletics Lag on Environmental Front," *NCAA News,* 25 August 2009, http://web1.ncaa.org/web_files/NCAANewsArchive/2009/Division+I/study%2Bstays%2Bcollege%2Bathletics%2Blags%2Bon%2Benvironmental%2Bfront_08_24_09_ncaa_news.html.

22. Jack Copeland, "From Little Things Big Things Grow," *NCAA Champion Magazine* (Summer 2008): 62, www.ncaachampionmagazine.org/search.html?ic=b5768d6e&sc=green%20team.

23. Jeff Orleans, phone interview with DeLongpré Johnston, 5 October 2009.

24. For information on STARS, see http://stars.aashe.org.

25. Paul Rowland, phone interview with DeLongpré Johnston, 29 September 2009.

26. News release: "Yale Presents Sustainable Athletics Goals at NCAA Convention," Yale University Office of Public Affairs, 22 January 2009, http://opa.yale.edu/news/article.aspx?id=6346.

27. Anna Prizzia, telephone interview with DeLongpré Johnston, 29 September 2009.

28. Whitewave Foods, Inc., a major organic foods producer, has renewed and increased its sponsorship of Ralphie's Green Stampede. Toyota and Frito-Lay are also sponsoring.

Chapter 17. The Impact of Sustainability on Institutional Quality Assurance and Accreditation

The author wishes to acknowledge her longtime Unity College (Unity, Maine) colleagues who in the 1990s introduced her to an institution of higher learning that embraced the unquestionable tenets of being good stewards of the earth. Dr. John Craig, longtime dean of admissions, masterfully incorporated in his outreach to prospective students and their families, at a time when it was not fashionable or politically correct, the environmentally conscious values and norms that guided student life at Unity in a way that in retrospect was visionary and strategic. Nearly three thousand miles away, in Olympia, Washington, Evergreen State College is one of several institutions in the Northwest that also embraces environmental stewardship in its programs and in the life of the institution. President Les Purce voices the importance of environmental sustainability by focusing on environmental security as critical to ensuring a more just world. Clearly, these are only two of the many institutions throughout the country that are engaging in good environmental stewardship and following good practices in leading the way toward greater sustainability for our nation and globally.

1. James Thompson, *Organizations in Action* (New York: John Wiley & Sons, 1967).

2. Charles T. Royer, "The State and Local Political Environment of Urban Health," *Washington Public Health* (Fall 2000): 3–4.

3. Society for College and University Planning, Sustainability in Higher Education, 2009, www.scup.org.

4. Text of the American College & University Presidents' Climate Commitment, 2007, www.presidentsclimatecommitment.org/about/commitment.

5. Ibid.

6. Association for the Advancement of Sustainability in Higher Education, "Sustainability Tracking, Assessment and Rating System for Colleges and Universities, Guide to Pilot Phase One," February 2008, www.aashe.org/documents/stars/STARS_Pilot_Phase_One_Guide.pdf.

7. Aspen Institute, "Beyond Grey Stripes: Preparing MBAs for Social and Environmental Stewardship," Business and Society Program, Aspen Institute, New York, 2005.

8. Michael Crow, "Sustainability: An Organizing Principle for Colleges and Universities," *University Business* (June 2007): 58–59.

9. Ibid.

10. See the UNESCO Education for Sustainable Development website, www.unesco.org/en/esd.

Chapter 18. Green Legal

Epigraph. Cited at www.RachelCarson.org.

1. Telephone conversation between Nilda Mesa and Arlene Lieberman, 21 September 2010.

2. Joshua Secunda, "An Experiment in 21st Century Enforcement: EPA–New England's Integrated Compliance Strategy for Colleges and Universities" (paper presented at the annual meeting of National Association of College and University Attorneys, Boston, June 2002).

3. Michael D. Sermersheim and Ronald Baylor, "Environmental Audits and Inspections: What Every University (and University Attorney) Should Know" (paper presented at the annual meeting of the National Association of College and University Attorneys, June 2004).

4. See www.epa.gov/lawsregs/laws.index.html.

5. See www.epa.gov.

6. Sermersheim and Baylor, "Environmental Audits and Inspections."

7. Secunda, "An Experiment in 21st Century Enforcement."

8. Martin Meehan, in a conversation with James Martin and James Samels, 4 October 2010.

9. Sermersheim and Baylor, "Environmental Audits and Inspections."

10. Ibid.

11. Brian Rogers, in an email to James Martin, 8 October 2010.

Chapter 19. Conclusion

1. Anthony Cortese, telephone interview with James Martin, 11 February 2010.

2. Burton R. Clark, "Delineating the Character of the Entrepreneurial University," in *On Higher Education: Selected Writings, 1956–2006* (Baltimore: Johns Hopkins University Press, 2008), 501.

3. Ibid., 501, 503, 506–7.

4. Anthony Cortese, email to James Martin, 6 August 2010.

5. Ibid.

6. Bill McKibben, "We're Hot as Hell and We're Not Going to Take It Any More," *TomDispatch.com,* 4 August 2010, www.tomdispatch.com/archive/175281.

7. Ibid.

8. Michael Brune, "Oil Spill Changes Everything," *CNN.com,* 1 May 2010, http://articles.cnn.com/2010-05-01/opinion/brune.oil.spill.danger_1_drilling-oil-spill-offshore?_s=PM:OPINION.

9. Glenn Cummings, "The Leadership Factor: Implementing Sustainability in Higher Education," 6 July 2010, http://secondnaturebos.wordpress.com/2010/07/06/the-leadership-factor-implementing-sustainability-in-higher-education.

10. Betty Klinck, "Find a Green College: Check!" *USA Today,* 20 April 2010.

11. John Tallmadge, as quoted in "What Is Creative Sustainability?" *Hawk and Handsaw: Journal of Creative Sustainability* 2 (2009): 4.

12. McKibben, "We're Hot as Hell."

13. Laura Matson, email to James Martin, 6 August 2009.

14. Ibid.

15. Howard Sacks, email to James Martin, 11 August 2009.

16. Mission statements taken from respective college websites: www.greenmtn.edu, www.coa.edu, www.northland.edu, and www.prescott.edu.

17. Mickey Z., "10 Ways to Promote Green Campus Activism," *Planet Green,* 6 August 2009, http://planetgreen.discovery.com/work-connect/promote-green-campus-activism.html.

18. Mitchell Thomashow, email to James Martin, 4 August 2009.

19. Quoted in Ben Eisen, "Sustainable for a Year," *Inside Higher Ed,* 10 July 2009, www.insidehighered.com/news/2009/07/10/sustainability.

20. Clark, "Delineating the Character," 500–501.

21. Peggy F. Barlett and Geoffrey W. Chase, eds., *Sustainability on Campus: Stories and Strategies for Change* (Cambridge, MA: MIT Press, 2004), 13.

22. Robert Weygand, email to James Martin, 12 August 2009.

23. Laura Matson, email to James Martin, 10 August 2009.

24. Stephen J. Nelson, "College Presidents and the Road to Success and Failure," *Bridgewater Review* 20, no. 1 (June 2009): 5.

25. Biographical information taken from the Ball State University website, www.bsu.edu.

26. Jo Ann Gora, email to James Martin, 6 August 2007.

27. David J. Skorton, "President's Climate Commitment," 23 February 2007, www.cornell.edu/statements/2007/20070223-presidents-climate-commitment.cfm.

28. Scott Carlson, "Cost and Red Tape Hamper Colleges' Efforts to Go Green," *Chronicle of Higher Education,* 11 April 2008, http://chronicle.com/article/CostRed-Tape-Hamper/14129.

29. Meredith Swinehart, "The Commons on Kennedy Mall: A Case Study in Green Building," August 2007, www.scu.edu/ethics/practicing/focusareas/environmental_ethics/green-building.html.

30. "Sidney J. Watson Arena Garners LEED Certification," 27 July 2009, http://athletics.bowdoin.edu/sports/general/20090727.

31. Cynthia Pollock Shea, telephone conversation with James Martin, 5 August 2008.

32. James Duderstadt, *Intercollegiate Athletics and the American University* (Ann Arbor: University of Michigan Press, 2000), 267.

33. Robert Weygand, email to James Martin, 12 August 2009.

34. Dave Hannon and Michelle Lang, "The Green Athlete," *New England Sports,* June 2008, 17.

35. Howard Greene and Matthew Greene, "Sustainable Admissions," *University Business* 11, no. 7 (July 2008): 57.

36. Editors, "Center for Rural Sustainable Development," *The Bell Tower,* University of Maine–Fort Kent, Spring 2008, 2.

37. Sara Hart, "Living Proof," *Continental Airways Magazine,* March 2009, 54–55.

38. Alice Good, "Nevada Home Grown," *Nevada Silver and Blue,* University of Nevada–Reno, Summer 2009, 30.

39. Greene and Greene, "Sustainable Admissions," 58.

40. Mickey Z., "10 Ways to Promote Green Campus Activism."

41. David S. Bernstein, "Generation Green," *Boston Phoenix,* 8 May 2009, 12.

42. John Cloud, "Synthetic Authenticity," *Time,* 24 March 2008, 53.

43. Tom de Castella, "Deep Thinkers," *Financial Times,* 21–22 November 2009, 1.

44. Wendy Koch, "'Eaarth' Author: Young People Can Save the World," *USA Today,* 20 April 2010, 7-D.

45. Davis Bookhart, email to James Martin, 26 August 2010.

46. Aurora Lang Winslade, email to James Martin, 19 August 2010.
47. Mary Jo Maydew, email to James Martin, 4 July 2010.
48. Paul Rowland, email to James Martin, 16 August 2010.
49. David E. Shi, "The Silver Lining in Forced Frugality," *Hawk and Handsaw: The Journal of Creative Sustainability* 3 (2010): 9.

Selected Bibliography

ACPA's Presidential Taskforce on Sustainability. "Change Agent Abilities Required to Help Create a Sustainable Future." www.myacpa.org/task-force/sustainability/docs/Change_Agent_Skills.pdf.

American College & University Presidents' Climate Commitment. www.presidentsclimatecommitment.org/html/commitment.php.

Anderson, Ray C. "Editorial: Earth Day, Then and Now." *Sustainability: The Journal of Record* 3, no. 2 (April 2010): 73–74.

Aspen Institute. "Beyond Grey Stripes: Preparing MBAs for Social and Environmental Stewardship." Business and Society Program, Aspen Institute, New York, 2005.

Association for the Advancement of Sustainability in Higher Education. "Higher Education Sustainability Officer Position and Salary Survey." January 2008. www.aashe.org/documents/resources/pdf/sustainability_officer_survey_2008.pdf.

———. STARS 1.0 Technical Manual, 2010. www.aashe.org/files/documents/STARS/STARS_1.0.1_Technical_Manual.pdf.

———. "Sustainability Tracking, Assessment and Rating System for Colleges and Universities, Guide to Pilot Phase One." February 2008. www.aashe.org/documents/stars/STARS_Pilot_Phase_One_Guide.pdf.

Atlee, Tom, with Rosa Zubizarreta. *The Tao of Democracy: Using Co-Intelligence to Create a World That Works for All.* North Charleston, SC: Imprint Books, 2003.

Bardaglio, Peter, and Andrea Putman. *Boldly Sustainable: Hope and Opportunity for Higher Education in the Age of Climate Change.* Washington, DC: National Association of College and University Business Officers, 2009.

Barlett, Peggy F. "Reason and Reenchantment in Cultural Change: Sustainability in Higher Education." *Current Anthropology* 49, no. 6 (2008): 1077–98.

———"Reconnecting with Place: Faculty and the Piedmont Project at Emory University." In *Urban Place: Reconnecting with the Natural World,* edited by Peggy F. Barlett, 39–60. Cambridge, MA: MIT Press, 2005.

Barlett, Peggy F., and Geoffrey W. Chase, eds. *Sustainability on Campus: Stories and Strategies for Change.* Cambridge: MIT Press, 2004.

Barlett, Peggy F., and Arri Eisen, "The Piedmont Project at Emory University." In *Teaching Sustainability at Universities,* edited by W. L. Filho, 61–77. Frankfurt: Peter Lang, 2002.

Barlett, Peggy F., and Anne Rappaport. "Long-Term Impacts of Faculty Development Programs: The Experience of TELI and Piedmont." *College Teaching* 57, no. 2 (2009): 73–82.

Barlow, Ben. *Financing Sustainability on Campus.* Washington, DC: National Association of College and University Business Officers, 2009.

Barnes, Niles. "AASHE Passes New Milestones in Time for Earth Day." AASHE Campus Sustainability Perspectives Blog, 21 April 2010. www.aashe.org/blog/aashe-passes-new-milestones-time-earth-day.

Belch, H. A., M. E. Wilson, and N. W. Dunkel. "Cultures of Success: Recruiting and Retaining New Live-In Residence Life Professionals." *Journal of College Student Development* 27, no. 2 (2009): 176–93.

Bleizeffer, Dustin. "Wyo Coal Miners Dig Record Volume in 2008." *Wyoming Energy Journal: Coal* (2009): 6–9.

Blewitt, John, and Cedric Cullingford, eds. *The Sustainability Curriculum: The Challenge for Education.* London: Earthscan, 2004.

Bok, Derek Curtis. *Universities and the Future of America.* Durham, NC: Duke University Press, 1990.

Bowers, Chet. *The Culture of Denial: Why the Environmental Movement Needs a Strategy for Reforming Universities and Public Schools.* Albany: State University of New York Press, 1997.

———. *Educating for an Ecologically Sustainable Culture: Rethinking Moral Education, Creativity, Intelligence, and Other Modern Orthodoxies.* Albany: State University of New York Press, 1995.

Braskamp, Larry A., and Jon F. Wergin. "Forming New Social Partnerships." In *The Responsive University: Restructuring for High Performance,* edited by William G. Tierney, 62–71. Baltimore: Johns Hopkins University Press, 1998.

Brook, Daniel. "It's Way Too Easy Being Green: The Decidedly Dupable System for Rating a Building's Greenness." *Slate,* 26 December 2007. www.slate.com/id/2180862.

———. "LEED Compliance Not Required for Designing Green Buildings: Constructing Buildings to the LEED Standard Can Conserve Energy and Materials—or Be Exploited for Promotional Gain." *Scientific American,* October 2008. www.scientificamerican.com/article.cfm?id=leed-compliance-not-required.

Bruntland, Gro Harlem, ed. *Report of the World Commission on Environment and Development: Our Common Future.* New York: United Nations, 1987. www.un-documents.net/wced-ocf.htm.

Camblin, L. D. Jr., and J. A. Steger. "Rethinking Faculty Development." *Journal of Higher Education* 39, no. 1 (2000): 1–18.

Chase, Geoffrey W., and Paul Rowland. "The Ponderosa Project: Infusing Sustainability in the Curriculum." In *Sustainability on Campus: Stories and Strategies for Change,* edited by Peggy F. Barlett and Geoffrey W. Chase, 91–106. Cambridge, MA: MIT Press, 2004.

Chesser, Mike. "Leadership and Learning: Driving toward Unprecedented Excellence." *Energybiz* 5, no. 5 (September-October 2008): 24–25.

Chrislip, David. *The Collaborative Leadership Fieldbook: A Guide for Citizens and Civic Leaders*. San Francisco: Jossey-Bass, 2002.

Clark, Burton R. "Delineating the Character of the Entrepreneurial University." In *On Higher Education: Selected Writings, 1956–2006*. Baltimore: Johns Hopkins University Press, 2008.

Colby, Anne, Thomas Ehrlich, Elizabeth Beaumont, and Jason Stephens. *Educating Citizens: Preparing America's Undergraduates for Lives of Moral and Civic Responsibility*. San Francisco, CA: Jossey-Bass, 2003.

Collett, Jonathon, and Stephen Karakashian, eds. *Greening the College Curriculum: A Guide to Environmental Teaching in the Liberal Arts: A Project of the Rainforest Alliance*. Washington, DC: Island Press, 1996.

Conscious Wave, Inc. "LOHAS Background." Lifestyles of Health and Sustainability (LOHAS) Online. www.lohas.com/about.html.

Corcoran, Peter Blaze, and Arjen E. J. Wals, eds. *Higher Education and the Challenge of Sustainability: Problematics, Promise, and Practice*. Dordrecht: Kluwer Academic Publishers, 2004.

Cortese, Anthony. "The Core Mission of Higher Education: Creating a Thriving, Civil and Sustainable Society." Paper presented at the ACPA Sustainability Institute, Harvard University, Cambridge, MA, 13 June 2000.

Cortese, Anthony, Georges Dyer, and Michelle Dyer. "Leading Profound Change: A Resource for Presidents and Chancellors of the ACUPCC." Boston, MA: ACUPCC, July 2009. www.presidentsclimatecommitment.org/files/documents/Leading_Profound_Change.pdf.

Crow, Michael. "Sustainability: An Organizing Principle for Colleges and Universities." *University Business* (June 2007): 58–59.

———. "Towards Institutional Innovations in America's Colleges and Universities." *Trusteeship* 3, no. 18 (May/June 2010). www.agb.org/trusteeship/2010/mayjune/toward-institutional-innovation-america%E2%80%99s-colleges-and-universities.

Cummings, Glenn. "The Leadership Factor: Implementing Sustainability in Higher Education." 6 July 2010. http://secondnaturebos.wordpress.com/2010/07/06/the-leadership-factor-implementing-sustainability-in-higher-education.

Cunz, Derek. "Integrated Delivery: Forget About the Contract, Focus on the Process." *Design Intelligence,* July 7, 2009. www.di.net/articles/archive/3100.

Curtis, Wayne. "A Cautionary Tale." *Preservation* 60, no. 1 (Jan/Feb 2008): 19–24.

Dautremont-Smith, Julian. "How Do Campus Sustainability Initiatives Affect College Submissions?" AASHE Campus Sustainability Perspectives Blog, 2 March 2009. www.aashe.org/blog/how-do-campus-sustainability-initiatives-affect-college-admissions.

———. "Ten Tips for Developing a Sustainability Game Plan for Athletic Departments." AASHE Campus Sustainability Perspectives Blog, 30 July 2009. www.aashe.org/blog/ten-tips-developing-sustainability-game-plan-athletic-departments.

DeLind, Laura B., and Terry Link. "Place as the Nexus of a Sustainable Future: A Course for All of Us." In *Sustainability on Campus: Stories and Strategies for Change,* edited by Peggy F. Barlett and Geoffrey W. Chase, 121–37. Cambridge, MA: MIT Press, 2004.

Deninger, Lynn, and John Swift. "Integrated Design: A Sustainable Mindset for Residence Halls." *Journal of College and University Housing* 36, no. 1 (2009): 48–71.

Dobson, Andrew, and Derek Bell, eds. *Environmental Citizenship*. Cambridge, MA: MIT Press, 2006.

"Domestic Energy Policy: Our Economic Future." Panel, Jackson Hole Policy Institute 2008. Senior Executive Energy Summit, Jackson Hole, Wyoming, November 12, 2008.

Duderstadt, James. *Intercollegiate Athletics and the American University.* Ann Arbor: University of Michigan Press, 2000.

Dumanoski, Dianne. *The End of the Long Summer: Why We Must Remake Our Civilization to Survive on a Volatile Earth*. New York: Crown Publishers, 2009.

The Earth Charter Initiative. *The Earth Charter.* www.earthcharterinaction.org.

Edwards, Andres R. *The Sustainability Revolution: Portrait of a Paradigm Shift.* Gabriola Island, BC: New Society Publishers, 2005.

Eisen, Arri, and Peggy F. Barlett. "The Piedmont Project: Fostering Faculty Development toward Sustainability." *Journal of Environmental Education* 38, no. 1 (2006): 25–38.

Eisen, Arri, Anne Hall, Tong Soon Lee, and Jack Zupko. "Teaching Water: Connecting across Disciplines and into Daily Life to Address Complex Societal Issues." *College Teaching* 57, no. 2 (Spring 2009): 99–104.

Epstein, Marc J. *Making Sustainability Work*. Sheffield, UK: Greenleaf Publishing Ltd., 2008.

Feldbaum, Mindy, with Hollyce States. *Going Green: The Vital Role of Community Colleges in Building a Sustainable Future and Green Workforce*. 2008. www.greenforall.org/resources/going-green-the-vital-role-of-community-colleges-in-building-a-sustainable-future-and-green-workforce.html.

Frederiksen, Chuck F. "A Brief History of Collegiate Housing." In *Student Housing and Residential Life,* edited by Roger B. Winston Jr. and Scott Anchors. San Francisco: Jossey-Bass, 1993.

Frumkin, Howard, Lawrence Frank, and Richard Jackson. *Urban Sprawl and Public Health: Designing, Planning, and Building for Healthy Communities*. Washington, DC: Island Press, 2004.

Garreau, Joel. *Radical Evolution: The Promise and Peril of Enhancing Our Minds, Our Bodies—and What It Means to Be Human*. New York: Doubleday, 2005.

Graham, Hugh Davis, and Nancy A. Diamond. *The Rise of American Research Universities: Elites and Challengers in the Postwar Era*. Baltimore: Johns Hopkins University Press, 1997.

Greene, Howard, and Matthew Greene. "Sustainable Admissions." *University Business* 11, no. 7 (July 2008): 57–58.

Hart, Stuart. *Capitalism at the Crossroads: Aligning Business, Earth, and Humanity,* 2nd ed. Upper Saddle River, NJ: Wharton School Publishing, 2007.

Hart Research Associates, "Raising the Bar: Employers' Views on College Learning in the Wake of the Economic Downturn: A Survey among Employers Conducted on Behalf of the Association of American Colleges and Universities," 20 January 2010. www.aacu.org/leap/documents/2009_EmployerSurvey.pdf.

Havens, Greg, Perry Chapman, and Bryan Irwin. "The Role of Sustainability in Campus Planning: A New England University Builds on the Land Grant Tradition." *New England Journal of Higher Education* 23, no. 2 (Fall 2008): 28–29.

Hawken, Paul. *Blessed Unrest: How the Largest Social Movement in History Is Restoring Grace, Justice, and Beauty to the World.* New York: Penguin, 2008.

Hayano, Delissa. "Guarding the Viability of Coal & Coal-Fired Power Plants: A Road Map for Wyoming's Cradle to Grave Regulation of Geologic CO_2 Sequestration." *Wyoming Law Review* 9 (2009): 139–41.

Higher Education Associations Sustainability Consortium (HEASC). "Call for a System for Assessing and Comparing Progress in Campus Sustainability." www.aashe.org/files/documents/STARS/HEASCcall.pdf.

Hondale, George. *How Context Matters: Linking Environmental Policy to People and Place.* West Hartford, CT: Kumarian Press, 1999.

Jackson, Tim. *Prosperity without Growth: Economics for a Finite Planet.* London: Earthscan Publisher, 2009.

Jones, Kristy M., and Julian L. Keniry. "National Trends in Sustainability Performance: Lessons for Facilities Leaders." *Facilities Manager* (March/April 2009): 46–49.

Jucker, Rolf. "'Sustainability? Never heard of it!': Some Basics We Shouldn't Ignore When Engaging in Education for Sustainability." *International Journal of Sustainability in Higher Education* 3, no. 1 (2002): 8–18.

Kammen, Daniel M. "The Rise of Renewable Energy." *Scientific American* 295, no. 3 (September 2006): 60–69.

Kaplan, Rachel, Stephen Kaplan, and Robert L. Ryan. *With People in Mind: Design and Management of Everyday Nature.* Washington DC: Island Press, 1998.

Kincheloe, Joe L., *Critical Pedagogy.* 2nd ed. New York: Peter Lang, 2008.

Korten, David C. *The Great Turning: From Empire to Earth Community.* Bloomfield, CT: Kumarian Press, 2006.

Langdon, Davis. "The Cost of Green Revisited: Reexamining the Feasibility and Cost Impact of Sustainable Design in the Light of Increased Market Adoption." July 2007. www.davislangdon.com/USA/Research/ResearchFinder/2007-The-Cost-of-Green-Revisited.

Maniates, Michael F. "Individualization: Plant a Tree, Buy a Bike, Save the World?" *Global Environmental Politics* 1, no. 3 (August 2001): 31–52.

Marten, Gerald G. *Human Ecology: Basic Concepts for Sustainable Development.* London: Earthscan Publications Ltd., 2001.

McDonough, Bill, and Michael Braungart. *Cradle to Cradle: Remaking the Way We Make Things.* New York: North Point Press, 2002.

McKenzie-Mohr, Doug, and William Smith. *Fostering Sustainable Behavior: An Introduction to Community-Based Social Marketing.* Gabriola Island, BC: New Society Publishers, 1999.

McSherry, Mark. 2009 Collegiate Athletic Department Sustainability Survey Report, June 2009. www.aashe.org/files/documents/resources/2009-Collegiate-Athletic-Department-Sustainability-Survey.pdf.

Meadows, Donella H., Dennis L. Meadows, Jorgen Randers, and William W. Behrens III. *The Limits to Growth.* White River Junction, VT: Chelsea Green Publishing Company, 2004.

Menand, Louis. *The Market Place of Ideas.* New York: W.W. Norton, 2010.

M'Gonigle, Michael, and Justine Starke. 2006. *Planet U: Sustaining the World, Reinventing the University.* Gabriola Island, BC: New Society Publishers, 2006.

Murphy, Kyle C. "Evaluating the Sustainability Tracking, Assessment and Rating System (STARS) at the Evergreen State College." Master's thesis, Evergreen State College, 2009.

Naditz, Alan. "The Green MBA: From Campus to Corporate." *Sustainability* 1, no. 3 (2008): 178–82.

National Environmental Education Foundation. "The Business Case for Environmental and Sustainability Employee Education" (February 2010). www.neefusa.org/business/index.htm.

Nelson, Stephen J. "College Presidents and the Road to Success and Failure." *Bridgewater Review* 20, no. 1 (June 2009): 3–6.

Noddings, Nel, ed. *Educating Citizens for Global Awareness.* New York: Teachers College Press, 2005.

Orr, David W. *Earth in Mind: On Education, Environment, and the Human Prospect.* Washington, DC: Island Press, 1994.

———. *Ecological Literacy: Education and the Transition to a Postmodern World.* Albany: State University of New York Press, 1992.

———. "Green Residence Halls Are Here: Current Trends in Sustainable Campus Housing." *Journal of College and University Student Housing* 36, no. 1 (2009): 10–23.

Owen, Ann L., and Julio Videras. "Trust, Cooperation, and Implementation of Sustainability Programs: The Case of Local Agenda 21." *Ecological Economics* 68 (December 2008): 259–72.

Palmer, Parker J. *The Courage to Teach: Exploring the Inner Landscape of a Teacher's Life.* San Francisco, CA: Jossey-Bass, 1998.

Patterson, Kerry, Joseph Grenny, Ron McMillan, and Al Switzler. *Crucial Conversations: Tools for Talking When the Stakes Are High.* Hightstown, NJ: McGraw-Hill, 2002.

Pelletier, Stephen. "Sustainability: What Is the Trustee's Stake?" *Trusteeship* 16, no. 5 (Sept/Oct 2008): 8–14.

Process of Preparation of the Environmental Perspective to the Year 2000 and Beyond. UN General Assembly Resolution 38/161, 19 December 1983. www.un.org/documents/ga/res/38/a38r161.htm.

Pryor, John H., et al. *The American Freshman: National Norms Fall 2008.* Los Angeles: Higher Education Research Institute, Graduate School of Education and Information Studies, University of California, Los Angeles, 2009.

Putman, Andrea, and Michael Philips. *The Business Case for Renewable Energy: A Guide to Colleges and Universities.* Alexandria, VA: APPA, 2006.

Rappaport, Ann, and Sarah Hammond Creighton. *Degrees That Matter: Climate Change and the University.* Cambridge, MA: MIT Press, 2007.

Report of the World Commission on Environment and Development, UN General Assembly Resolution 42/187, 11 December 1987. www.un.org/documents/ga/res/42/ares42-187.htm.

Rose, Rodney. *Buildings . . . The Gifts That Keep on Taking.* Alexandria, VA: APPA Center for Facilities Research, 2007.

Royer, Charles T. "The State and Local Political Environment of Urban Health." *Washington Public Health* (Fall 2000): 3–4.

Savitz, Andrew W., and Karl Weber. *The Triple Bottom Line: How Today's Best-Run Companies Are Achieving Economic, Social, and Environmental Success—and How You Can Too.* San Francisco: Jossey-Bass, 2006.

Scigliano, J. A. "Faculty Development: Issues and Directions." *Peabody Journal of Education* 55, no. 2 (1978): 152–60.

Selby, David. "Degrees of Denial: As Global Heating Happens, Should We Be Educating for Sustainable Development or Sustainable Contraction?" In *Talking Truth, Confronting Power,* edited by Jerome Satterthwaite, Michael Watts, and Heather Piper, 17–34. Sterling, VA: Trentham, 2008.

Shellenberger, Michael, and Ted Nordhaus. "The Death of Environmentalism: Global Warming Politics in a Post-Environmental World." Essay. Environmental Grantmakers Association, 29 April 2004.

Shi, David E. "The Silver Lining in Forced Frugality." *Hawk and Handsaw: The Journal of Creative Sustainability* 3 (2010): 9.

Sigmon, William L. "The Lure of Ultra-Supercritical: Exploring the Future of Coal-Burning." *Energybiz* 5, no. 5 (September-October 2008): 90–91.

Skorton, David J. "President's Climate Commitment," 23 February 2007. www.cornell.edu/statements/2007/20070223-presidents-climate-commitment.cfm.

Sorcinelli, M. D., A. E. Austin, P. L. Eddy, and A. L. Beach. *Creating the Future of Faculty Development: Learning from the Past, Understanding the Present.* Bolton, MA: Anker Publishing Company, 2006.

Steffen, Alex. "It's Not Just Carbon, Stupid." *Wired* 16, no. 6 (June 2008): 165.

Stiglitz, Joseph. *Freefall: America, Free Markets, and the Sinking of the World Economy.* New York: W.W. Norton, 2010.

Stonich, Susan C. "Producing Food for Export: Environmental Quality and Social Justice Implications of Shrimp Mariculture in Honduras." In *Who Pays the Price: The Socio-Cultural Context of Environmental Crisis,* edited by Barbara Rose Johnston, 110–20. Washington, DC: Island Press, 1996.

Svanström, Magdalena, Francisco Lozano, and Debra Rowe. "Learning Outcomes for Sustainable Development in Higher Education." *International Journal of Sustainability in Higher Education* 9, no. 3 (2008): 339–51.

Thomashow, Mitchell. *Bringing the Biosphere Home: Learning to Perceive Global Environmental Change.* Cambridge, MA: MIT Press, 2002.

———. *Ecological Identity: Becoming a Reflective Environmentalist.* Cambridge, MA: MIT Press, 1995.

Thompson, James. *Organizations in Action.* New York: John Wiley & Sons, 1967.

Torres-Antonini, Maruja, and Norbert W. Dunkel. "Green Residence Halls Are Here: Current Trends in Sustainable Campus Housing." *Journal of College and University Student Housing* 36, no. 1 (2009): 10–23.

Uhl, Christopher. *Developing Ecological Consciousness: Path to a Sustainable World.* New York: Rowman and Littlefield, 2004.

University of North Carolina Tomorrow Commission, Final Report, December 2007. www.northcarolina.edu/nctomorrow/reports/commission/Final_Report.pdf.

University of Wyoming Campus Sustainability Committee. "Greenhouse Gas Emissions Inventory for the University of Wyoming, Update Fiscal Year 2008." http://uwyo.edu/sustainability/pcc.asp.

University of Wyoming School of Energy Resources. *First Annual Report.* Prepared for the Joint Minerals, Business and Economic Development Interim Committee, Joint Appropriations Interim Committee, and the Joint Education Interim Committee, October 2006. www.uwyo.edu/sersupport/docs/2006report.pdf.

———. *2008 Annual Report.* Prepared for the Joint Minerals, Business and Economic Development Interim Committee, Joint Appropriations Interim Committee, and the Joint Education Interim Committee, October 2008. www.uwyo.edu/sersupport/docs/SER%20Annual%20Rpt%202008%20Final.pdf.

U.S. Green Building Council. "U.S. Green Building Council Strategic Plan 2009–2013." www.usgbc.org/DisplayPage.aspx?CMSPageID=1877.

Veysey, Laurence R. *The Emergence of the American University.* Chicago: University of Chicago Press, 1965.

Viederman, Stephen. "Can Universities Contribute to Sustainable Development?" In *Inside and Out: Universities and Education for Sustainable Development,* edited by Linda Silka and Robert Forrant, 17–29. Amityville, NY: Baywood Publishing Company, 2006.

Wackernagel, Mathis, and William Rees. *Our Ecological Footprint: Reducing Human Impact on the Earth.* Gabriola Island, BC: New Society Publishers, 1996.

Wergin, J. F., E. J. Mason, and P. J. Munson. "The Practice of Faculty Development: An Experience-Derived Model." *Journal of Higher Education* 47, no. 3 (1976): 289–308.

"What Is Creative Sustainability?" *Hawk and Handsaw: The Journal of Creative Sustainability* 2 (2009): 4.

Wheatley, Margaret J. *Leadership and the New Science: Discovering Order in a Chaotic World.* San Francisco: Berrett-Koehler Publishers, 1999.

Wilson, Mel. "Investing in the Future." *Communications Review* 12, no. 1 (2007): 13–17.

Wuppertal Institute. "Prism of Sustainability." www.foeeurope.org/sustainability/sustain/t-content-prism.htm.

Contributors

JAMES MARTIN has been a member of the Mount Ida College faculty since 1979. Now a professor of English, he served for over fifteen years as the college's vice president for academic affairs and provost. A Methodist minister, he was awarded a Fulbright Fellowship to study mergers in the University of London system. He has been academic vice president of The Education Alliance since 1986.

With his writing and consulting partner, James E. Samels, Martin has coauthored four previous books available from the Johns Hopkins University Press: *Merging Colleges for Mutual Growth* (1994), *First among Equals: The Role of the Chief Academic Officer* (1997), *Presidential Transition in Higher Education: Managing Leadership Change* (2005), and *Turnaround: Leading Stressed Colleges and Universities to Excellence* (2009). He co-writes a column on college and university issues, "Future Shock," for *University Business*. Martin and Samels also co-hosted the nation's first television talk program on higher education issues, *Future Shock in Higher Education,* from 1994 to 1999 on the Massachusetts Corporation for Educational Telecommunications (MCET) satellite learning network. A graduate of Colby College (A.B.) and Boston University (M.Div. and Ph.D.), Martin has co-written articles for the *Chronicle of Higher Education,* the *London Times,* the *Christian Science Monitor,* the *Boston Globe, Trusteeship, CASE Currents,* and *Planning for Higher Education.*

JAMES E. SAMELS is the founder and CEO of both The Education Alliance and the Samels Group, a full-service higher education consulting firm. He is also the founding partner of Samels Associates, a law firm serving independent and public colleges, universities, and nonprofit and for-profit higher education organizations. Samels has served on the faculties of the University of Massachusetts and Bentley College and as a guest lecturer at Boston University and Harvard University. Prior to his appointment at the University of Massachusetts, Samels served as the deputy and acting state comptroller in Massachusetts, special assistant attorney general, Massachusetts Community College counsel, and general counsel to the Massachusetts Board of Regents.

Samels holds a bachelor's degree in political science, a master's degree in public administration, a juris doctor degree, and a doctor of education degree. He has written and co-written a number of scholarly articles, monographs, and opinion editorials appearing in the *Chronicle of Higher Education, AGB Trusteeship,* the *Christian Science Monitor,* the *London Guardian,* the *Boston Globe,* the *Boston Herald, Boston Business Journal, Journal of Higher Education Management,* and *Planning for Higher Education.* He is the coauthor, with James Martin, of *Merging Colleges for Mutual Growth* (1994), *First among Equals: The Role of the Chief Academic Officer* (1997), *Presidential Transition in Higher Education: Managing Leadership Change* (2005), and *Turnaround: Leading Stressed Colleges and Universities to Excellence* (2009), all from the Johns Hopkins University Press. Samels has previously consulted on projects and presented research papers at universities, colleges, schools, and ministries of education in China, Canada, Great Britain, France, Korea, Sweden, Thailand, and Turkey.

MICHAEL A. BAER is a vice president at Isaacson, Miller, specializing in higher education. Prior to joining Isaacson, Miller in 2005, he served as senior vice president at the American Council on Education (ACE) where he oversaw all ACE programming.

PEGGY BARLETT is the Goodrich C. White Professor of Anthropology at Emory University. She is a specialist in agricultural anthropology and serves on the National Research Council Board on Agriculture and Natural Resources. A leader in bringing Emory to its current commitment to a sustainable future, Peggy has focused in recent years mainly on faculty development, curriculum, and sustainable food. She and Geoffrey Chase have edited a volume of narratives on the challenges of transforming higher education, *Sustainability on Campus: Stories and Strategies for Change* (2004).

Co-founder of the Piedmont Project at Emory, the longest-running curriculum development program for sustainability in the nation, she now leads workshops on faculty engagement with sustainability around the country through the Association for the Advancement of Sustainability in Higher Education (AASHE). Her work highlights the importance of place, and her edited book, *Urban Place: Reconnecting with the Natural World* (2005), presents research on the mental and physical health impacts of nature contact. Her current interests focus on strategies to maintain resilience and creativity in cultural change and practices of restoration. Barlett holds a bachelor's degree from Grinnell College and a doctoral degree from Columbia University, both in anthropology.

DAVIS BOOKHART was hired in 2006 to create a program and coordinate the various sustainability efforts at Johns Hopkins University. That effort has evolved into the university's Office of Sustainability. As director, Bookhart leads the team in facilitating projects that reduce the negative environmental impacts of the university while promoting sustainability through collaboration between divisions and involvement in the larger Baltimore community.

Bookhart chairs the Johns Hopkins Sustainability Committee, a multidivisional group convened to focus on strategies and vision for the university. In 2008, Bookhart was appointed commissioner to the Baltimore City Sustainability Commission; he also serves on the Advisory Council for the Association for the Advancement of Sustainability in Higher Education (AASHE) and is a member of the editorial board of *Sustainability: The Journal of Record.*

Bookhart holds a master's degree in international affairs from the Fletcher School of Law and Diplomacy at Tufts University and a master's in American literature from the University of North Carolina at Wilmington. He co-founded and remains president of Charm21—Clean and Healthy Air through Renewables in Maryland—a nonprofit group advocating the use of renewable fuels and resources in the Baltimore region.

THOMAS BUCHANAN's career in higher education has spanned more than thirty-five years, as a student, teacher, and administrator. A native of New York, Buchanan attended the State University of New York at Cortland, where he obtained his undergraduate degree in 1973. He earned a master's of science from the University of Wyoming and a Ph.D. from the Institute for Environmental Studies at the University of Illinois at Urbana-Champaign. After completing his doctorate, Buchanan returned as an assistant professor in the Department of Geography at the University of Wyoming. Over the next thirty years, he became full professor, and he has held various administrative positions, including department head, associate dean of the College of Arts and Sciences, and vice president for academic affairs.

In 2005, he was appointed the 23rd president of the University of Wyoming. As UW president, Buchanan's priorities for the university have included excellence in academics, promoting access to higher education in Wyoming, and enhancing economic and workforce development in Wyoming. Buchanan is the recipient of numerous awards recognizing excellence in teaching and administration, including the Ellbogen Meritorious Classroom Teaching award and the Seibold Professorship in the College of Arts and Sciences. He serves on the governing boards of the Mountain West Athletic Conference, the Western Interstate Commission for Higher Education, and the Western Cooperative for Educational Telecommunications.

SCOTT CARLSON, who joined the *Chronicle of Higher Education* in 1999, writes about facilities, energy, architecture, and sustainability. With Lawrence Biemiller, he runs the Buildings & Grounds blog and assembles the annual architecture issue. A former technology reporter, he is a host of the *Tech Therapy* podcast, which also features the technology consultant and cohost Warren Arbogast. In 2006, Carlson broke the news that the FBI was interested in rifling the papers of the journalist Jack Anderson, held at George Washington University. His article led to national coverage and Senate hearings and helped Carlson win first prize for beat reporting from the National Education Writers Association that year. One of his earlier pieces, "The Deserted Library," published in 2001, generated national attention for its coverage of the place of libraries in an Internet age.

Before coming to the *Chronicle,* Carlson worked at the *Star Tribune* and *City Pages,* both in Minneapolis, and at *City Paper* in Baltimore. He has also written for national magazines like the *Utne Reader* and *Dwell,* and he is a contributing writer for *Urbanite,* an award-winning magazine that focuses on city living and sustainability in Baltimore, where he lives. He is a graduate of the University of Minnesota–Twin Cities.

GEOFFREY CHASE, dean of undergraduate studies and director of the Center for Regional Sustainability at San Diego State University, attended Ohio Wesleyan University, where he received a B.A. in English. He also holds an M.A.T. from Miami University (Ohio) and an A.M. in English from Boston College. After receiving his Ph.D. in American literature from the University of Wisconsin–Madison, he taught for eleven years in the School of Interdisciplinary Studies at Miami University of Ohio. While at Miami, he served as a Fulbright Scholar in Turku, Finland.

Chase joined Northern Arizona University in 1992 as the director of English Composition. While at Northern Arizona University, Chase also served as chair of the English Department, dean of Liberal Studies, and as the associate provost for Undergraduate Studies. At NAU, he revamped the composition curriculum to give it an environmental focus and became a leader of the Ponderosa Project, a faculty development project aimed at helping faculty throughout the university integrate issues of environmental sustainability into their courses. The Ponderosa Project has become a model that has been introduced to faculty on more than 175 campuses in the United States and Canada.

In 2004 he co-edited, with Peggy Barlett, *Sustainability on Campus: Stories and Strategies for Change.* He currently serves as board chair for the Association for the Advancement of Sustainability in Higher Education (AASHE). He has also served on the executive committee for the American Conference of Academic Deans (ACAD), and as co-chair for the Proposal Review Committee for the Western Association of Schools and Colleges (WASC).

ANTHONY D. CORTESE, SC.D., is co-founder, with Senator John Kerry (D-MA) and Teresa Heinz, and president of Second Nature, a nonprofit organization with a mission to develop the national capacity to make healthy, just, and sustainable action a foundation of all learning and practice in higher education.

He is also a co-organizer of the American College & University Presidents' Climate Commitment and co-founder of the Association for the Advancement of Sustainability in Higher Education. He is co-founder and co-coordinator of the Higher Education Association Sustainability Consortium and a consultant to higher education, industry, and nonprofit organizations on institutionalization of sustainability principles and programs. He is actively involved with the presidents and chancellors of colleges and universities and several higher education associations that represent presidents, trustees, business officers, facilities managers, planners, purchasing agents, housing officers, sports program administrators, and student affairs officers to promote sustainable design, education, planning, operation, purchasing, and community collaboration in higher education.

Cortese was formerly commissioner of the Massachusetts Department of Environmental Protection. He was the first dean of Environmental Programs at Tufts University and founded the award-winning Tufts Environmental Literacy Institute in 1989 that helped integrate environmental and sustainability perspectives in more than 175 courses.

He also organized the effort that resulted in the internationally acclaimed Talloires Declaration of University Leaders for a Sustainable Future in 1990 and which has now been signed by more than 365 presidents and chancellors in more than fifty countries.

Cortese was a founding member of the board of directors of the Natural Step US and of the Environmental Business Council of New England. He is a trustee of Green Mountain College, Woodrow Wilson Fellow for higher education, a fellow of the American Association for the Advancement of Science, a member of the EPA Science Advisory Board, and a member of the President's Council on Sustainable Development's Education Task Force. Cortese has B.S. and M.S. degrees from Tufts University in civil and environmental engineering, a Doctor of Science in Environmental Health from the Harvard School of Public Health and an honorary Ph.D. from Allegheny College.

DEDEE DELONGPRÉ JOHNSTON is the director of sustainability at Wake Forest University. She holds a bachelor's degree in business administration from the University of Southern California with a concentration in entrepreneurial studies and a master's of business administration with an emphasis in sustainable management from the Presidio Graduate School in San Francisco. Prior to her role at Wake Forest, DeLongpré Johnston served as director of the University of Florida's Office of Sustainability where she pioneered the development of the university's first campus-wide strategic plan for sustainability. She was named one of the 10 Innovators of the Year in 2007 by *Florida Trend* magazine and was featured in a cover story in the October 2008 issue of *Sustainability: The Journal of Record.*

DeLongpré Johnston has almost twenty years of experience in nonprofit management in the areas of education, sustainability, and environmental conservation. She served as the executive director of the nonprofit Sustainable Alachua County (FL) and was the U.S. program director for Fauna and Flora International, the world's longest-standing conservation NGO. She is currently vice chair of the board of the Association for the Advancement of Sustainability in Higher Education (AASHE).

LYNNE DENINGER serves as vice president for Cannon Design in Boston, Massachusetts. With twenty years of experience planning and designing various project types, she has developed a wide-ranging expertise in the design of university architecture. Strongly committed to green design, Deninger is a LEED-accredited professional with experience managing interdisciplinary teams of professionals responsible for establishing integrated green strategies. A graduate of the Rhode Island School of Design, she maintains an active presence in the Boston Society of Architects and is a regular presenter at national industry conferences.

Deninger has specialized in the programming, planning, and design for student residence facilities for almost a decade, and she has played a significant role in the planning and design of student residences for the following institutions: Worcester Polytechnic Institute, University of Wisconsin–Whitewater, University of South Carolina, Michigan State University, George Mason University, and the University of North Carolina at Chapel Hill.

NORBERT W. DUNKEL is the assistant vice president and director of housing and residence education at the University of Florida. His primary responsibilities include serving as chief housing officer for 10,000 students and family members, an operation with 750 employees and with an operating budget of $40 million. Dunkel also recently served as president of the Association of College and University Housing Officers—International (ACUHO-I) with more than nine hundred member institutions from seventeen countries. He has authored or edited ten books and monographs and more than forty chapters or articles on various aspects of campus housing. He has served as a consultant to more than twenty universities and colleges and has twice testified before congressional committees.

SANDRA ELMAN is the president of the Northwest Commission on Colleges and Universities (NWCCU) in Redmond, Washington. Dr. Elman served as chair of the Council of Regional Accrediting Commissions (CRAC), which is composed of the directors and chairs of the seven regional accrediting commissions, from 2003 to 2006. Prior to assuming the position of the president of NWCCU in 1996, Dr. Elman was the associate director of the Commission on Institutions of Higher Education of the New England Association of Schools and Colleges. Before joining regional accreditation, Elman held a variety of administrative and faculty positions at the John McCormack Institute of Public Affairs at the University of Massachusetts, the University of Maryland, and the University of California–Berkeley. She has published extensively in the fields of public policy and higher education and is coauthor of *New Priorities for the University: Educating Competent Individuals for Applied Knowledge and Societal Needs.*

She is an adjunct faculty member at Oregon State University and serves as an evaluator for international quality assurance agencies, including for the Center for Accreditation and Quality Assurance of Swiss Universities. Elman is a past chair of the Board of Trustees of Unity College in Unity, Maine. Elman received her B.A. degree in history and political science from Hunter College in New York and her M.A. and Ph.D. in policy, planning, and administration from the University of California–Berkeley. She is also a 2005 graduate of the Department of Defense National Security Seminar, U.S. Army War College.

TARA EVANS, a native of Wyoming, graduated from the University of Wyoming with a bachelor's of science in molecular biology in 2002. She earned her doctor of jurisprudence from the University of Wyoming in 2006. After completing her doctorate, Evans was a judicial clerk for the Honorable Bruce E. Kasold in the United States Court of Appeals for Veteran's Claims in Washington, D.C., and also worked as assistant attorney general for Wyoming, practicing water and natural resource law. In 2008, Evans joined the legal staff of the University of Wyoming part-time, working as a special assistant to the president of the University of Wyoming. She now works full time as associate general counsel. Evans has been an adjunct instructor for the Department of Criminal Justice and the Haub School of Environment and Natural Resources at the University of Wyoming. She is currently an adjunct instructor at the University of Wyoming College of Law.

Evans is also a member of the Campus Sustainability Committee, which is charged by the president to identify sustainable initiatives that will support the American College & University Presidents' Climate Commitment and highlight the university's leadership in energy resources.

RICK FAIRBANKS, as provost and vice president of academic affairs at Northland College, Fairbanks, led the creation of an integrated, interdisciplinary general education curriculum, restructured academic programs and majors, supervised the establishment of a coherent and integrated co-curriculum, completed a successful diversity plan for faculty hiring, and initiated and led several successful grant proposals, including a Title III proposal totaling nearly $2 million.

Prior to his time at Northland, Fairbanks served as the associate dean for humanities at St. Olaf College in Northfield, Minnesota, where he provided oversight of staff planning, hiring, budget management, and capital and noncapital expenditures. In addition, Fairbanks oversaw facilities planning and administrative support for department chairs, program directors, and eighty faculty members in eight departments and four interdisciplinary programs. While at St. Olaf, Fairbanks also served as a professor of philosophy and the department's chair. Before joining the faculty at St. Olaf, Fairbanks taught philosophy at the University of Chicago, Augustana College, the University of Minnesota, and Concordia College in Moorhead, Minnesota.

A native of Minot, North Dakota, and a graduate of Concordia College with a B.A. in philosophy and classics, Fairbanks went on to earn his M.A. in religion at the University of Chicago, and his Ph.D. in philosophy at the University of Minnesota.

TIMOTHY FARNHAM is the Leslie and Sarah Miller Director of the Center for the Environment at Mount Holyoke College. Previously, Farnham served as associate professor and director of the Undergraduate Program in Environmental Studies at the University of Nevada, Las Vegas.

SEAN FARRELL is a Managing Associate at Isaacson, Miller.

JO ANN GORA became Ball State's fourteenth president in 2004 and has led the development of the university's Education Redefined Strategic Plan, the cornerstone of which is making immersive learning opportunities available to every student. The plan is the basis of the university's capital campaign, Ball State Bold.

Since her arrival and stretching through 2012, the university is spending approximately $418 million on major construction and renovation projects. President Gora also has led Ball State's sustainability efforts, including its geothermal energy project, the largest of its kind in the nation. President Gora is a member of the Association of Governing Boards' Council of Presidents. She chairs the Mid-American Conference Presidents' Council and co-chairs the Central Indiana Corporate Partnership.

ROBERT J. KOESTER is a professor of architecture at Ball State University. He teaches design-for-sustainability studios, sustainability seminars, vital signs courses, and co-teaches the *Daylectric*™ studio—daylighting and electrical lighting integration in architectural design.

He is founding director of the Center for Energy Research/Education/Service (CERES), a university-level unit; founding chair of BSU's Council on the Environment (COTE), a clearinghouse for campus-wide sustainability; and founding co-chair of the Greening of the Campus (GOC) conference series. He is also a founding member of the Board of Directors of the Association for the Advancement of Sustainability in Higher Education (AASHE).

TERRY LINK was appointed executive director of the Greater Lansing Food Bank in Michigan in 2009. Prior to that, during more than two decades of service at Michigan State University, he served as founding director of the Office of Campus Sustainability for eight years and was chair of the university's Committee for a Sustainable Campus. He was also an adjunct faculty member in the Bailey Scholars Program.

LAURA MATSON was the original STARS technical developer for AASHE. Previously, she had worked in various capacities for Portland, Oregon's Office of Sustainable Development. Matson holds an honors degree in economics from Lewis & Clark College, where she received the Worldly Philosopher Award. While at Lewis & Clark, she was active in sustainability advocacy and programs. Her research appears in *Frontiers in Ecological Economic Theory and Application*. She is now a full-time graduate student pursuing an M.S. in natural resources and environment and an M.A. in urban planning at the University of Michigan.

MARY JO MAYDEW has been the vice president of finance and administration at Mount Holyoke College for more than twenty years. Previously, she had been assistant treasurer at Cornell University. During her time at Mount Holyoke, Maydew has also served terms as board chair for the National Association of College and University Business Officers (NACUBO) and president of the board for the Eastern Association of College and University Business Officers (EACUBO).

DAVE NEWPORT is director of the University of Colorado at Boulder Environmental Center. Founded on Earth Day 1970, it is the nation's first and largest student-funded center of its kind. The Environmental Center's staff members operate higher education's original recycling program as well as alternative transportation programs, energy and climate conservation programs, sustainable food, environmental justice, and environmentally preferred purchasing initiatives. Newport also chairs the university's Carbon Neutrality Working Group and is a faculty associate in the Environmental Studies Department, where he teaches a course in Carbon Neutrality Planning for Higher Education.

Newport is secretary of the Board of Directors of the Association for the Advancement of Sustainability in Higher Education (AASHE) and a member of the three-person steering committee for AASHE's campus sustainability ratings system (STARS). Prior to his position in Colorado, Newport served as sustainability director at the University of Florida, project manager of Florida's first climate neutrality assessment project, and coauthor of the first comprehensive sustainability assessment of a college campus performed to international business standards.

DEBRA ROWE has been a pioneering voice and leading advocate in sustainability education and initiatives for more than thirty years. She is president of the U.S. Partnership for Education for Sustainable Development, which convenes members of the business, education, government, and faith sectors to catalyze sustainability initiatives. At Oakland Community College in Michigan, Dr. Rowe is a professor of energy management, renewable energy technology, and psychology, focusing on harnessing energy from renewable sources. She also teaches energy management and renewable energies in an online format. She created an Energy Awareness Center at OCC and has hosted many conferences and customized trainings on energy and sustainable design practices.

As a consultant to a national consortium of community colleges, the Partnership for Environmental Technology Education (PETE), Rowe created a model energy management degree design for community colleges, funded by the U.S. Department of Energy. Colleges around the country are currently using the materials to develop their own programs. Rowe is a consultant to more than twenty colleges, helping them incorporate energy and/or sustainability into their institutions' operations and curricula. In addition, she is the energy and sustainability consultant to the National Science Foundation–funded National Science Database Library. She is working on the creation of the national eERL collection (electronic Environmental Resources Library). Before coming to OCC to teach, Dr. Rowe owned a renewable energy and energy management company, conducting energy audits and designing/installing energy efficiency and renewable energy systems. She is on the education division board of the American Solar Energy Society.

Rowe received her Ph.D. from the School of Business Administration at the University of Michigan in 1991. She received her M.A. in psychology in 1989 and her M.B.A. in 1988 from the University of Michigan in Ann Arbor. Her bachelor's degree is from Yale University in 1977.

HOWARD L. SACKS is the National Endowment for the Humanities Distinguished Teaching Professor of Sociology at Kenyon College, where he directs the Rural Life Center. From 2002 to 2009, he served on Kenyon's senior administration as provost and senior advisor to the president. Sacks currently serves on the Governor's Ohio Food Policy Advisory Council. He operates a sheep farm with his wife.

LEITH SHARP has almost twenty years of experience in greening universities around the world. She has consulted and presented to more than one hundred organizations and is on the governing committees and editorial boards of numerous organizations, including the Association for the Advancement of Sustainability in Higher Education and the *International Journal of Sustainability in Higher Education*. Sharp has received numerous awards for her work, including a Churchill Fellowship and Young Australian of the Year, NSW Environment Category.

From 2000 to 2008, Sharp was the founding director of Harvard University's Green Campus Initiative and led the creation of the largest green campus organization in the world, taking Harvard to the forefront as a global leader in campus sustainability. Under her leadership, Harvard achieved more than fifty LEED building projects, instituted a $12-million revolving loan fund that achieved an average return on investment of 30 percent, and implemented wide-scale engagement in occupant behavioral change, onsite renewable energy projects, GHG reduction commitments, alternative fuels, green cleaning, and environmental purchasing. Sharp is currently engaged in a variety of writing, teaching, speaking, and consulting activities. She has an ongoing affiliation as a visiting scholar with the Harvard School of Public Health and continues to teach organizational change management for sustainability and green building design through Harvard's Extension School. She holds a bachelor's degree in engineering (environmental) from the University of New South Wales (Australia) and a master's degree in education (human development and psychology) from Harvard University.

CINDY POLLOCK SHEA launched the Sustainability Office at the University of North Carolina, Chapel Hill, in April 2001. She has researched, analyzed, written about, and communicated sustainability themes at international research institutes such as Worldwatch and the International Institute for Sustainable Development and has worked as a foreign correspondent, web scribe, and community organizer. She holds degrees in economics, international relations, and environmental studies from the University of Wisconsin–Madison and is a LEED-accredited professional.

MARY SPILDE was appointed president of Lane Community College in 2001. She joined Lane in 1995 as vice president for instructional services and then vice president for instruction and student services. Prior to that, she spent fifteen years at Linn-Benton Community College in Albany, Oregon, in a variety of positions, including dean of business, health, and training. She earned a bachelor's degree in business and social systems and a law degree from the University of Edinburgh, Scotland. She completed a master's in adult education and the doctorate in post-secondary education at Oregon State University.

CYNTHIA THOMASHOW is executive director of the Center for Environmental Education and a faculty member at Unity College.

MITCHELL THOMASHOW has been president of Unity College since 2006. He is the author of *Ecological Identity: Becoming a Reflective Environmentalist* (1995) and *Bringing the Biosphere Home* (2001).

JUDY WALTON is the membership and outreach director of AASHE. She was the founding executive director of AASHE and the founding director of Education for Sustainability Western Network (EFS West). Her interests in sustainability and higher education are long-standing. As a faculty member at Humboldt State University during the early 2000s, she played a key role in campus sustainability efforts. Prior to that she worked for a green building consultancy in Washington State, when "green building" was a new field. Walton has delivered presentations to campuses and businesses across the United States and Canada, assisted campuses with strategic planning, and organized national and international events on sustainability and higher education. She holds a Ph.D. in geography from Syracuse University, an M.A. in geography from San Diego State University, and a B.A. in political science with a minor in economics from the University of California–San Diego.

AURORA LANG WINSLADE is the Sustainability Director at the University of California–Santa Cruz. In that position, she has specialized in strategic and operational planning, conflict mediation, program development, information architecture for web development, and green building and operations.

Prior to her present position, Winslade was co-founder and statewide co-coordinator of the California Education for Sustainable Living Program, a project on five University of California campuses and one community college campus sponsored by the California Student Sustainability Coalition. In 2003, she served as a marketing intern for the California Certified Organic Farmers Association. Winslade is a graduate of the University of California–Santa Cruz.

Index